Musikautomaten, Moden und Uniformen im Technik Museum Speyer

~•~

Automatic Musical Instruments, Fashion, and Uniforms in the Technik Museum Speyer

TECHNIK
MUSEUM
SPEYER

Inhalt - Contents

 Fotografie, Texte, Layout und CD-Mastering: Dr. Hans-Jürgen Schlicht Multimedia & Print, Sinsheim. Übersetzung: Dr. Hans-Jürgen Schlicht und Bianca Sowders. Unser herzlicher Dank gilt Gotthard Arnold, Helga Erbacher und Sina Hildebrand von der Fachstätte historischer Musikautomaten in Bad Schönborn für die tatkräftige Unterstützung.

ISBN 3-9809437-4-7

Das Museum

Die Technik Museen Sinsheim und Speyer beherbergen eine einzigartige Sammlung mechanischer Musikinstrumente, von der winzigen Spieldose bis zur hausgroßen Tanzorgel. In diesem Teil des Buches werden die wichtigsten Instrumente im Technik Museum Speyer vorgestellt. Die mechanischen Musikinstrumente sind in Speyer auf zwei Bereiche verteilt. Viele Großinstrumente, darunter die weltweit größte Welte Philharmonie Orgel sowie eine 12,5 Meter breite Mortier Tanzorgel, die breiteste Tanzorgel der Welt, befinden sich in der als Liller Halle bezeichneten Ausstellungshalle des Technik Museums. Die überwiegende Mehrzahl der Instrumente wird jedoch im Wilhelmsbau gezeigt, der nur wenige Schritte von der Liller Halle entfernt ist. Sein Wahrzeichen ist die vom Speyerer Bildhauer Wolf Spitzer aus Edelstahl geformte, 15 Meter hohe Großplastik „Orpheus".

Das Spektrum der Instrumente im Wilhelmsbau reicht von den Anfängen der Musikautomaten, wie den Serinetten und Flötenuhren des ausgehenden 18. Jahrhunderts, bis zu den perfektionierten Reproduktionsklavieren und automatischen Geigen des frühen 20. Jahrhunderts. Die meisten Instrumente sind voll spielfähig und können von den Besuchern in Gang gesetzt oder im Rahmen von Führungen in Aktion erlebt werden. Einen Eindruck von der Klangqualität gibt die diesem Buch beiliegende Audio-CD. Bei den Instrumenten ist vermerkt, wenn sich ein Musikbeispiel auf der CD befindet.

The Technik Museums Sinsheim and Speyer house a unique collection of automatic musical instruments ranging from tiny music boxes to dance organs as tall as a house. This section of the book presents the most important instruments shown in the Technik Museum Speyer are presented. In Speyer, the automatic musical instruments are shown in two areas. Many large instruments, among them the largest Welte Philharmonic Organ in the world as well as a Mortier dance organ, with a width of 12.5 m the widest dance organ in the world, are located in the Liller Halle, the large exhibition hall of the Technik Museum Speyer. The majority of the instruments, however, are displayed in the Wilhelmsbau, just a few steps away from the Liller Halle. Its landmark is the towering sculpture "Orpheus" with a height of 15 m, created from stainless steel by the local sculptor Wolf Spitzer.

The spectrum of instruments shown in the Wilhelmsbau ranges from the early automatic instruments like the serinettes and flute-playing clocks from the late 18th century to the perfected reproducing pianos and self-playing violins of the early 20th century. Most of the instruments are fully functional and can either be activated by the visitors or can be experienced during a guided tour. The audio CD in this book gives an impression of their sound quality. You will find a note at the instruments if a music example is included on the CD.

Der Eingang zum Wilhelmsbau mit der Großplastik „Orpheus". - The entrance to the Wilhelmsbau with the towering sculpture "Orpheus".

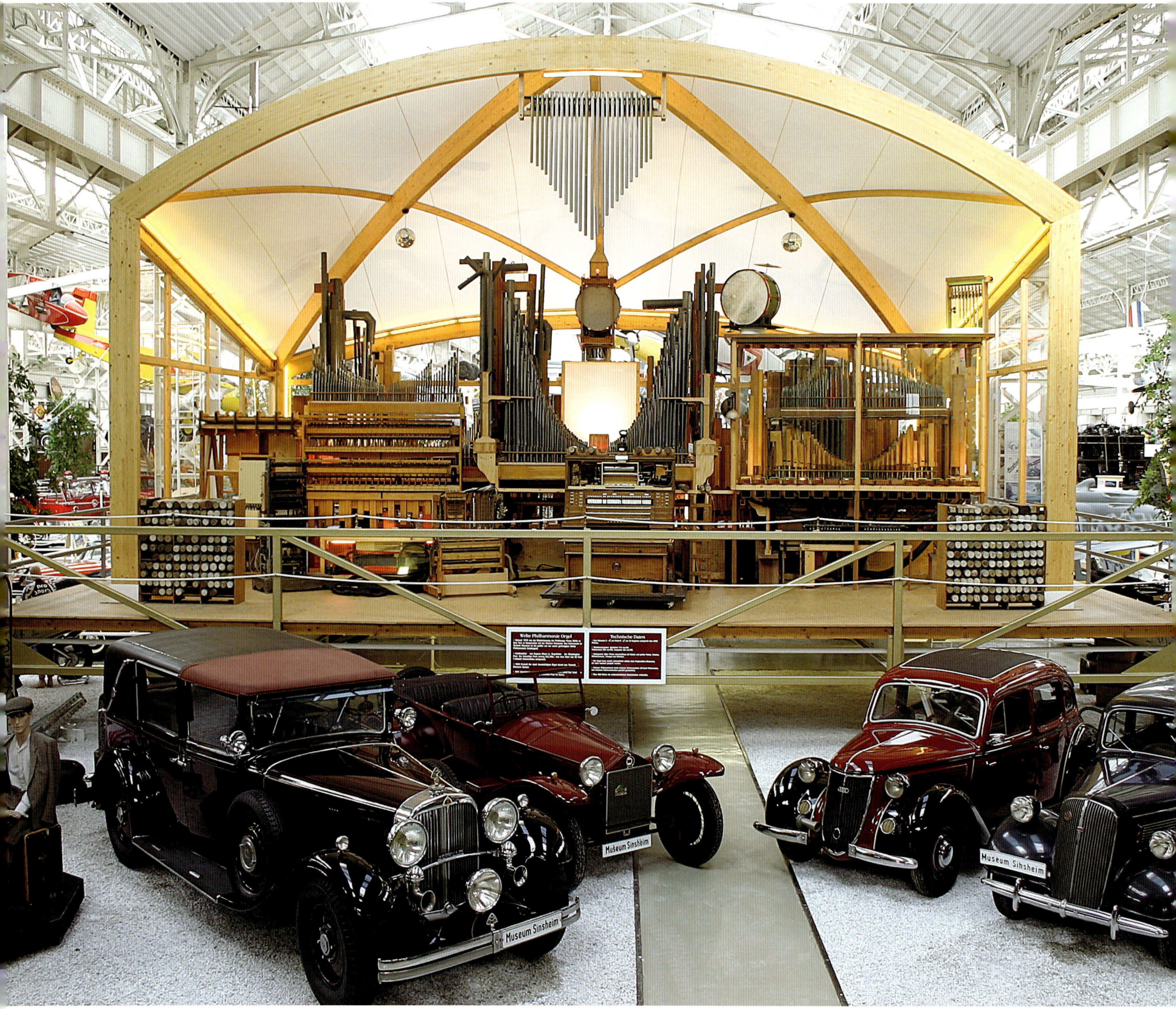
Welte Philharmonie Orgel
Technische Daten
Museum Sinsheim
Museum Sinsheim
Museum Sinsheim

Die Liller Halle

- Erbaut 1913 in Lesquin / Lille für die Firma Thomson, Houston
- Abgebaut während des 1. Weltkriegs durch deutsche Truppen
- Transport nach Speyer und Wiederaufbau für die Pfalz-Flugzeugwerke, die hier 2500 Flugzeuge fertigten
- Nach dem 1. Weltkrieg bis 1930 erfolgte eine Nutzung durch französische Truppen
- Von 1937 bis Anfang 1945 diente sie den Flugzeugwerken Saarpfalz als Werkstatt hauptsächlich für Militärflugzeuge
- Nach 1945 kam es erneut zu einer militärischen Nutzung durch französische Truppen
- 1984 Übergabe an das Bundesvermögensamt
- 1989 Übergabe an das Technik Museum Speyer
- Am 11. April 1991 Eröffnung als Museumshalle
- Die Halle hat eine Länge von 110 m, eine Breite von 60 m und eine Maximalhöhe von 15 m
- Gezeigt werden u.a. Oldtimer, Flugzeuge, Hubschrauber, Feuerwehrfahrzeuge, Lokomotiven und Musikautomaten

The Liller Halle

- Built in 1913 at Lesquin / Lille for the Thomson Co., Houston
- Dismantled by German troops during WW I
- Transported to Speyer and reconstructed for the aircraft builder Pfalz-Flugzeugwerke who built 2500 planes in this building
- After WW I the hall was used by the French Army until 1930
- From 1937 until early 1945 the aircraft builder Saarpfalz used it as a workshop specially for military planes
- After 1945 it was once more used by French troops
- 1984 it was handed over to the German authorities
- 1989 it was handed over to the Technik Museum Speyer
- On 11. April 1991 it was opened as an exhibition hall
- The hall has a length of 110 m, a width of 60 m, and a maximum height of 15 m
- On display are, among many other items, vintage cars, airplanes, helicopters, fire engines, locomotives, and automatic musical instruments

Links: Die Welte Philharmonie Orgel in der Liller Halle. Bilder rechts: Die Ausstellungsschwerpunkte in der Liller Halle sind Flugzeuge, Hubschrauber, Oldtimer, Lokomotiven und Feuerwehren.

Left: The Welte Philharmonic Organ in the Liller Hall. Pictures right: The major items shown in the Liller Hall are airplanes, helicopters, vintage cars, locomotives and fire engines.

CONCERT ORIGINAL.

Der Wilhelmsbau

- Erbaut 1917 vom Architekten Albert Speer sen. als Verwaltungsgebäude der Pfalz Flugzeugwerke
- Nach 1945 Nutzung durch französische Truppen
- 1984 Übergabe an das Bundesvermögensamt
- Aufgrund mangelnden Interesses an der Erhaltung des Gebäudes kommt es in den folgenden Jahren zu großen Zerstörungen durch Vandalismus
- 1989 Übergabe an das Technik Museum Speyer. Es folgt eine mehrjährige Totalsanierung.
- Am 20. November 1999 Eröffnung als Museumsgebäude
- Das renovierte Bauwerk hat eine Länge von 53,50 m und eine Breite von 17,05 m und bietet über 4000 m^2 Ausstellungsfläche auf vier Etagen
- Gezeigt werden mechanische Musikinstrumente, historische Moden und Accessoires, Juwelen, Puppen und Spielzeug, historische Waffen und Uniformen sowie Jagdtrophäen

The Wilhelmsbau

- Built by the architect Albert Speer Sr. in 1917 as an administrative building for the Pfalz Aircraft Company
- After 1945 it was used by French troops
- 1984 it was handed over to the German authorities
- Due to lack of interest in the preservation of the building, large parts were destroyed by vandalism
- In 1989, the building was handed over to the Technik Museum Speyer and restored over several years
- On 20. November 1999 it was opened as an exhibition building
- The renovated building has a length of 53,50 m and a width of 17,05 m and offers more than 4000 sqm exhibition space on four floors
- On exhibit are automatic musical instruments, historical fashion and accessories, jewels, dolls and toys, historical weapons and uniforms as well as hunting trophies

Links: Der „Braune Salon" im Wilhelmsbau. Rechts oben: Das „Rote Zimmer" mit der weltberühmten Hupfeld Phonoliszt-Violina. Rechts Mitte: Das 50er Jahre Zimmer - Ein Beispiel für die vielen Puppenarrangements mit historischen Moden im Wilhelmsbau. Rechts unten: Die Sammlung Winkler im Obergeschoss des Wilhelmsbau beinhaltet eine Vielzahl von historischen Uniformen, Waffen und Moden.

Left: The "Brown Salon" in the Wilhelmsbau. Top right: The "Red Room" with the world-famous Hupfeld Phonoliszt-Violina. Middle right: The "Fifties Room" - Just one example of the many arrangements with historical fashion decorated on life-like mannequin dolls in the Wilhelmsbau. Bottom right: The Winkler collection comprises a multitude of historical uniforms, weapons, and fashion.

Eine besondere Attraktion im Wilhelmsbau ist die Ausstellung historischer Moden und Accessoires. Eine kleine Auswahl zeigen die nächsten Seiten.

Bild oben: Tafelgesellschaft aus der Zeit um die Jahrhundertwende. Bild rechts: Kinderkleid aus Seide mit aufgestickten Blumen. Es entstand um 1860 und ist eines der ältesten und wertvollsten Stücke der Sammlung.

A particular attraction in the Wilhelmsbau is the exhibition of historical fashion and accessories. A brief selection is depicted on the next pages.

Picture top: Dinner party at the turn of the century. Picture right: Child dress made of silk with flower embroidery. It was made circa 1860 and is one of the oldest and most valuable items of the collection.

Viele der Moden sind auf Künstlerpuppen dekoriert, die so lebensecht wirken, dass man meint, sie müssten sich jeden Moment durch den Raum bewegen.

Many of the fashion garments are modeled on mannequin dolls, so life-like that you would expect to see them move around at any moment.

Die Sammlung historischer Moden im Wilhelmsbau umfasst zahllose kostbare und seltene Raritäten, wie sie in kaum einem anderen Museum zu sehen sind. Sie spannt einen weiten Bogen von der Gründerzeit der zweiten Hälfte des 19. Jahrhunderts über die sich daran anschließende Epoche des Jugendstils bis in die 1950er Jahre. Viele der Ausstellungsstücke waren beim Erwerb in einem problematischen Zustand und mussten aufwändig restauriert werden. Gefährdete Stoffpartien, Löcher und Risse wurden mit einem stützenden Gewebe unterlegt und das meist aus reiner Seide bestehende Innenfutter, das für Alterungsprozesse besonders anfällig ist, ersetzt. Schwierig war dabei die Beschaffung geeigneter Ersatzmaterialien, da diese den Originalstoffen so nah wie möglich kommen sollten. Arrangiert in liebevoll gestalteten Vitrinen mit Gegenständen des täglichen Alltags gibt die Modeausstellung im Wilhelmsbau einen faszinierenden Einblick in den Zeitgeist und die Lebensart der letzten 150 Jahre.

The historic fashion collection exhibited in the Wilhelmsbau comprises many rare and valuable items seldomly found in any other museum. It spans a range from the days of industrial expansion in the second half of the 19th century, over the ensuing Art Nouveau period, up to the 1950s. A great number of the exhibits were in poor condition when obtained and needed painstaking restoration. Fabric parts at risk, as well as holes and tears were lined with supporting material. The lining on the inside, which was mainly made of pure silk and especially subject to aging, was replaced. Obtaining adequate substitute materials was often challenging since the replacement fabrics were to match the originals as closely as possible. Carefully arranged in showcases together with many items of everyday's life, the historic fashion exhibition in the Wilhelmsbau offers a fascinating insight into the way of life of the last 150 years.

Gegenüberliegende Seite links: Tüllkleid mit Spitze (links) und Abendkleid aus Tüllspitze mit Pailletten bestickt (rechts), beide aus den 1950er Jahren. Gegenüberliegende Seite rechts: Tageskleider um 1905 aus Baumwollbatist mit Spitze und Bändchenborte eingefasst. Oben links: Balltoillette um 1893. Das zweiteilige Kleid aus schwerer Atlas-Seide ist mit Perlen und Metallfäden bestickt und mit hellgrünem Samt eingefasst. Die Taillenweite betrug nur 46 cm mit einschnürendem Korsett. Aufgrund des eingeschränkten Brustkorbes kam es beim Tanzen vielfach zu Ohnmachten bei den Damen. Oben rechts: Mit Perlen besticktes Nachmittagskleid um 1907 mit eingesetztem Schulterkragen.

Opposite page left: Tulle dress with lace (left) and evening gown made of tulle-lace with sequin-embroidery (right), both from the 1950s. Opposite page right: Informal dresses from around 1905, made of cambric, with lace- and ribbon trimming. Top left: Full ball gown from around 1893. The two-piece gown of heavy satin is adorned with embroidery of pearls and metallic threads and trimmed with light green velvet. Laced in, it was made to fit a waistline of merely 46 cm. As a consequence of the severely constricted chest and breathing capacity it was quite common for ladies to faint. Top right: Tea gown from around 1907 with pearl embroidery and collar set into the shoulders.

Links: Morgentoillette um 1900 mit einem Unterteil aus Seide und einem Überkleid aus Organza. Mitte: Abendkleid um 1900 aus reiner Seide mit edler Spitze (Südstaaten / USA). Rechts: Abendkleid um 1900 aus Bändchenspitze mit einem Schal aus Organza (Atlanta / Georgia, USA).

Left: Morning gown from around 1900 with a full-length silk slip under a dress made of organza. Middle: Pure silk evening gown from around 1900 with precious lace (Southern States / USA). Right: Ribbon-lace evening gown from around 1900 with organza collar (Atlanta, Georgia / USA).

Oben links: Vitrine mit Mode und Accessoires aus den 1950er Jahren. Oben rechts: Nachmittagskleid aus den 1920er Jahren gefertigt aus Seidentaft mit Strass-Stickerei (links). Abendrobe aus den 1930er Jahren aus reiner Seide mit Strass-Stickereien (rechts). Das Gewicht der Pailletten ließ die Kleider während des Tanzes zu den rhythmischen Bewegungen ihrer Trägerinnen mitschwingen. Den damaligen Modetänzen entsprechend bekamen sie den Namen „Shimmy-" oder „Charlestonkleider". Unten links: Sommerliches Promenadenkleid um 1896 aus Seidenchiffon eingefasst mit Bändchenkordeln und einem Unterrock aus raschelndem Seidentaft.

Top left: Display case with fashion and accessories from the 1950s. Top right: Tea gown from the 1920s made of silk-taffeta with paste embroidery (left). Pure silk evening gown from the 1930s with paste embroidery (right). While dancing, the very weight of the sequins made the dresses swing in rhythm with the wearer's movements, and in accordance with the dances popular at that time they were dubbed "Shimmy" or "Charleston dresses". Bottom left: Summer wear promenade dress from around 1896 made of silk-chiffon with ribbon-cord trimming and a petticoat of rustling silk-taffeta.

Was sind Musikautomaten?

Die wesentlichste Eigenschaft der mechanischen Musikinstrumente, die auch als Musikautomaten bezeichnet werden, ist, dass sie Musikstücke selbsttätig spielen können. Vor der Elektrifizierung erfolgte der Antrieb durch Gewichte, Federwerke oder von Hand. Ab 1890 setzte sich der Elektromotor durch.

Alle Musikautomaten verfügen über drei entscheidende Bauelemente, nämlich einen Tonträger, einen Antrieb und einen oder mehrere Klangerzeuger. Automaten, die viele unterschiedliche Klangerzeuger beinhalten, werden als Orchestrien bezeichnet, da sie die Klangfülle eines kleinen Orchesters besitzen. Üblicherweise unterscheidet man die Musikautomaten anhand des Tonträgers. Die wichtigsten Tonträger sind Stiftwalzen, perforierte Metallscheiben und gelochte Papierrollen bzw. Kartonbänder.

Die frühen Musikautomaten verwendeten Stiftwalzen als Tonträger. Das Bild links zeigt den Instrumententeil eines herrlichen Pfeifenorchestrions von Imhof & Mukle, das weiter hinten noch im Detail beschrieben wird. Es entstand 1861, als die Walzensteuerung bereits weit entwickelt war. Die verwendete Technik ist typisch für diese Instrumentenbauart.

Als Antrieb dient ein Gewicht, das von Hand nach oben gekurbelt werden muss. Dieses treibt sowohl die Stiftwalze als auch eine Pumpe im unteren Geräteteil an, das die Luft für die Pfeifen liefert. Später wurden bei Walzenorchestrien auch Federwerke oder elektrische Antriebe verwendet. Die Stifte werden durch Hebel abgetastet, die direkt oder indirekt die Klangerzeuger bzw. die Pfeifenventile betätigen.

What are Automatic Musical Instruments?

The most characteristic feature of automatic musical instruments is that they can play pieces of music on their own. Before electrification they were driven by weights, spring-wound clockworks or by hand. After 1890 the electro motor became dominant.

All automatic instruments contain three crucial parts: a drive, a sound carrier and one or more sound sources or instruments. Automatic instruments with many different sound sources are also referred to as orchestrions since they possess the sound volume of a small orchestra. Automatic instruments are usually distinguished by the type of sound carrier. The most widely used sound carriers are pinned barrels, perforated metal disks, and perforated cardboard and paper bands.

All early automatic instruments used pinned barrels as sound carriers. The picture on the left shows the instrument part of a beautiful pipe orchestrion made by Imhof & Mukle. It was built in 1861 when barrel technology was already quite advanced.

It is driven by a weight that had to be wound up by hand and drives both the pinned barrel and a pump in the lower part of the instrument that delivers the compressed air for the pipes. Later on barrel orchestrions were also driven by spring-wound clockworks or electro motors. The pins are sensed by levers which either directly or indirectly activate the sound sources or the pipe valves, respectively.

Die Steuerung des Imhof & Mukle Pfeifenorchestrion ist typisch für ein Walzenorchestrion. Das Gewicht ist an einem Seil (1) aufgehängt, und treibt beim Herabsinken die Stiftwalze (2) an. Über Hebelstößel (3) wird das Schöpfwerk im unteren Geräteteil betätigt, das Luft in eine Windlade einbläst. Ein Fliehkraftregler (4) regelt die Umdrehungsgeschwindigkeit. Die Abtastung der Stiftwalze erfolgt durch Hebel (5). Bei Orchestrien ohne Pfeifen entfällt das Schöpfwerk.

The controls of the Imhof & Mukle pipe orchestrion are typical for a barrel orchestrion. The pinned barrel (2) is driven by a weight that is attached to a rope (1). Via tappets (3) it also drives the pump in the lower part of the instrument that blows air into a reservoir. A centrifugal force regulator (4) controls the speed of rotation. The pinned barrel is sensed by levers (5). Orchestrions without pipes lack the pump.

ochmann's
Original
La Palom
Die Taube)
Mexikan
ches Lied

Stiftwalzen als Tonträger hatten mehrere Nachteile. Die Walzen waren teuer und ein Walzenwechsel recht umständlich. Auch war die Lauflänge der Musikstücke stark begrenzt. Dies konnte teilweise dadurch kompensiert werden, dass die Stifte spiralförmig eingeschlagen und die Walze beim Abspielen seitlich verschoben wurde, so dass ein Musikstück auf mehr als eine Walzenumdrehung verteilt werden konnte. Außerdem konnten mehrere Musikstücke nebeneinander eingeschlagen werden, so dass die Walze für ein neues Stück nicht ausgetauscht, sondern nur seitlich verschoben werden musste.

Trotzdem stand nur eine sehr eingeschränkte Musikauswahl zur Verfügung, was besonders in den Gasthäusern schnell Langeweile erzeugte. Gegen Ende des 19. Jahrhunderts kamen daher Instrumente auf den Markt, die mit einer Metallscheibe gesteuert wurden. Der Antrieb erfolgte meist durch einen Federaufzug oder durch Gewichte, nur ganz wenige Geräte besaßen bereits einen Elektromotor. In die Metallplatte wurden Löcher gestanzt, wodurch an der Rückseite Ausbuchtungen entstanden, die meist über Sternräder und Hebel die Klangerzeuger betätigten. Die einfachen Geräte verfügten meist nur über einen oder zwei Klangerzeuger und werden auch als Polyphone bezeichnet, um sie von den aufwändiger instrumentierten Orchestrien abzugrenzen. Aufgrund der leichten Austauschbarkeit der Platten werden die Polyphone manchmal auch als Vorläufer der Plattenspieler angesehen, was aufgrund der völlig unterschiedlichen Technik aber nicht gerechtfertigt ist.

Pinned barrels as sound carriers had several disadvantages. The barrels were quite expensive and changing them was very complicated. Also the length of a piece of music was rather limited. This could at least partially be compensated by a spiral arrangement of the pins and a lateral shifting of the barrel during play. In this way, a piece of music could be distributed over more than one barrel revolution. In addition one barrel could contain several pieces of music side by side. To play a new tune it was not necessary to change the barrel but only to shift it laterally.

Despite all these tricks the choice of music remained limited, which tended to become boring especially in pubs and guesthouses after a while. Therefore, automatic musical instruments that used perforated metal discs as sound carrier were developed towards the end of the 19th century. They are referred to as disc music boxes. They were mostly driven by spring-wound clockworks or weights, only very few instruments were already equipped with an electro motor. The metal discs were punched with holes causing bulges to appear on the backside which activated the sound sources via star-wheels and levers. Simpler instruments contained only one or two sound sources and are also called polyphones to distinguish them from the more sophisticated orchestrions. Because the discs were easy to change, polyphones are sometimes regarded as predecessors of modern record players. However, this comparison is not justified since the technology used by polyphones and record players is completely different.

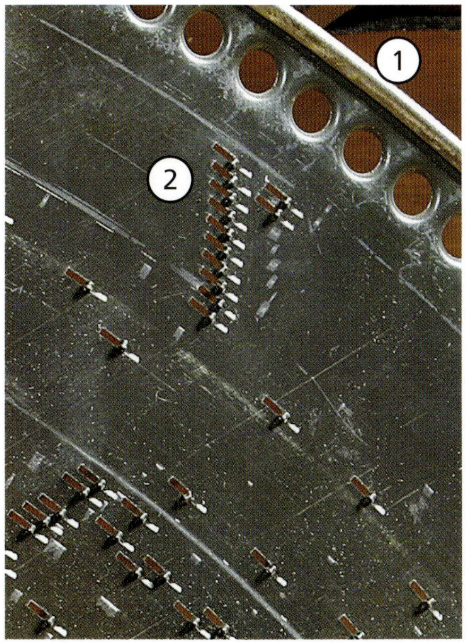

Die Steuerung eines Plattenautomaten am Beispiel des Lochmann 300 Concert Original. Das Bild ganz links zeigt die Rückseite einer Platte mit der Randlochung für den Antrieb (1) und den Metallhaken zur Auslösung der Klangerzeuger (2). Im Bild daneben die Antriebs- und Abtasteinheit im Gerät. Der Antrieb erfolgt über ein Sternrad (1), das in die Randlochung der Platte einrastet. Die Abtastung erfolgt ebenfalls über Sternräder, die die Klangerzeuger, z.B. hier einen Metallkamm (3), entweder direkt auslösen oder indirekt über Hebel betätigen (4).

The controls of a disc music box using the Lochmann 300 Concert Original as an example. The picture on the left shows the back of as disc with the peripheral holes for the drive (1) and the metal hooks that activate the sound sources (2). The picture beside it shows the drive and sensing components within the instrument. The disc is driven by a star wheel (1) which engages into the peripheral holes of the disc. Sensing of the disc also occurs by star wheels which activate the sound sources, for instance a metal comb (3), either directly or via levers (4).

Neubayrischer Ländler
Sammeljäger Marsch

Zur vollen Blüte gelangten die Musikautomaten durch die Verbesserung der Pneumatik zu Beginn des 20. Jahrhunderts. Das Prinzip ist einfach. In einem meist elektrisch angetriebenen Schöpfwerk wird je nach Steuerungstechnik ein Unterdruck, ein Überdruck oder beides erzeugt. Als Tonträger dient ein auf einer Rolle aufgewickeltes Papierband oder ein gefaltetes, perforiertes Kartonband. Das Papier- oder Kartonband wird über einen sogenannten Spieltisch geführt, in dem die Steuerleitungen der Pneumatikventile enden. In dem Moment, in dem eine Öffnung im Papierband den Lesekopf überquert, strömt Luft von außen ein und ändert dadurch die Druckverhältnisse im System. Hierdurch werden Steuerventile ausgelöst, die über einen pneumatischen Mechanismus die Klangerzeuger betätigen.

Nicht alle Instrumente, die perforierte Kartonbänder als Tonträger verwenden, werden pneumatisch gesteuert. Es wurden auch einfache Musikautomaten gebaut, bei denen die Kartonbänder, ähnlich wie bei den Plattenautomaten, mechanisch abgetastet wurden.

Alle im folgenden beschriebenen Musikautomaten basieren auf den hier dargestellten Prinzipien der Walzen-, Platten-, Papier- oder Kartonbandsteuerung.

Automatic musical instruments reached their climax towards the beginning of the 20th century due to improved pneumatic controls. The principle is simple. Positive or negative pressure, or both, depending on the type of control mechanism, are generated in a pump driven by an electro motor. A perforated paper band wound on a roll or a perforated cardboard folded zig-zag fashion into a compact pile are used as sound carriers. The paper or cardboard band is guided over a tracker bar where the control lines of the pneumatic valves end. When a perforation in the paper or cardboard band crosses one of the openings in the tracker bar, outside air streams into the system and changes the pressure ratios. As a result, the attached valves are switched and the sound sources are activated.

Not all instruments which use perforated cardboard bands as sound carrier have a pneumatic control system. There are also simple instruments which, like disc music boxes, sense the perforation mechanically.

All automatic musical instruments described in the following use pinned barrels, punched metal discs or perforated paper or cardboard bands as sound carriers as described in this introduction.

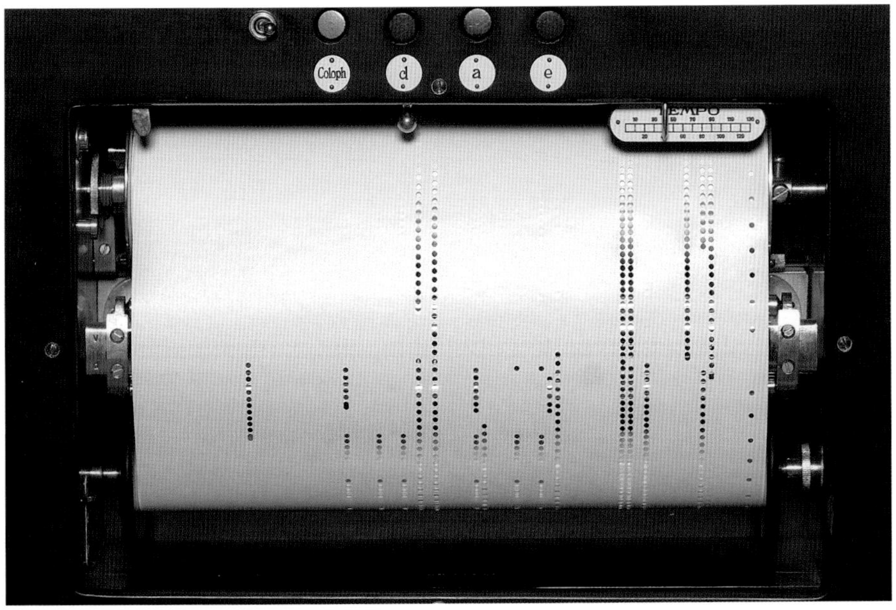

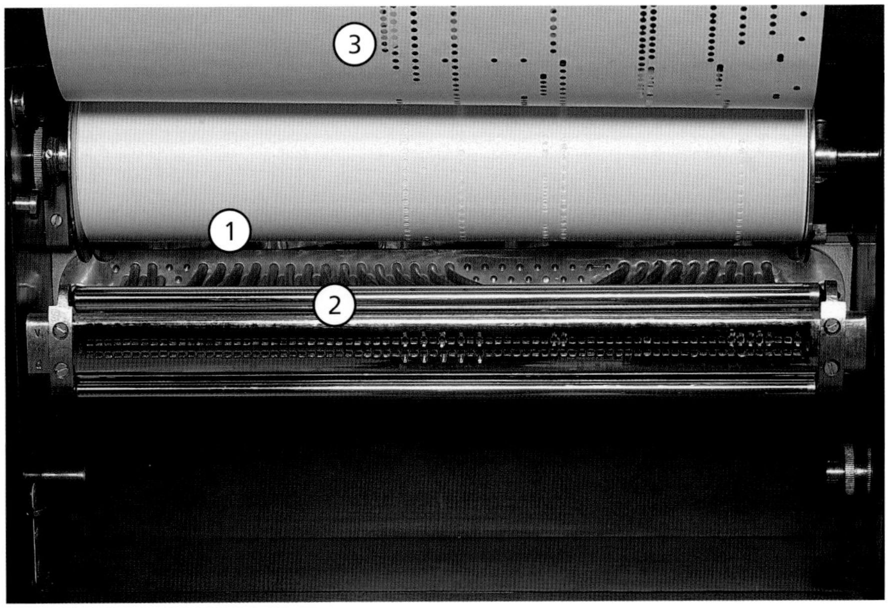

Gegenüberliegende Seite: Der Spieltisch des „Weber Erato" Orchestrions. Die Steuerung erfolgt durch ein perforiertes Papierband. Ersatzrollen können direkt im Gerät gelagert werden. Bilder oben: Ein typischer Spieltisch mit und ohne Papierrolle. Die Steuerleitungen der Pneumatikventile (1), deren Öffnungen im Spieltisch enden (2), sind deutlich zu erkennen. Der Luftdruck in den Leitungen wird durch die Perforation in der Papierrolle (3) geregelt. Die Klangerzeuger werden pneumatisch über Ventile betätigt, die durch die Druckänderungen geschaltet werden.

Opposite page: The tracker bar of the "Weber Erato" orchestrion. It is controlled by a perforated paper roll. Spare rolls can be stored directly inside the instrument. Pictures top: A typical tracker bar with and without the paper roll. The control lines of the pneumatic valves (1) which end in the tracker bar (2) are clearly visible. The pressure in the lines is controlled by the perforation in the paper roll (3). The sound sources are activated via pneumatic valves that are switched by the pressure changes.

Flöteninstrumente

Die ältesten Geräte, die im weiten Sinne als mechanische Musikinstrumente bezeichnet werden können, sind die Spielwerke von Uhren. Anfänglich handelte es sich um Glockenspiele oder um vom Psalter abgeleitete Harfenuhren, die aufgrund der geringen Klangvielfalt aber nur sehr einfache Melodien spielen konnten. Im 18. Jahrhundert erfreuten sich die Serinetten, auch als „Vogelorgeln" bezeichnet, einer großen Beliebtheit. Sie wurden u.a. verwendet, um Kanarienvögeln das Singen beizubringen. Ein Beispiel aus dem Wilhelmsbau zeigt die Abbildung unten.

Serinetten besitzen bereits alle charakteristischen Bauteile eines mechanischen Musikinstruments: einen Antrieb (Handkurbel), einen Tonträger (Stiftwalze) sowie Klangerzeuger (Pfeifen). Die Handkurbel treibt sowohl die Walze als auch das Schöpfwerk an. Die Abtastung der Stiftwalze erfolgt durch kleine Hebel, die jeweils mit einem Ventil verbunden sind. Wird ein Hebel durch einen Stift angehoben, öffnet sich das Ventil und die im Schöpfwerk verdichtete Luft kann in die zugehörige Pfeife strömen.

Die Verfeinerung dieses Funktionsprinzips führte schließlich zu den Flötenuhren, in die man zusätzlich zum Uhrenteil ein Flötenwerk einbaute. Als Antrieb dient das Aufzugsgewicht der Uhr. Durch die Verbesserung der Stiftwalzen und der Abtastmechanik konnten Melodien immer perfekter wiedergegeben werden, was dazu führte, dass selbst so berühmte Komponisten wie Mozart, Haydn, Händel oder Beethoven Stücke eigens für Flötenuhren komponierten. Die wertvollsten Meisterwerke fanden sich in den Schlössern der Kaiser und Könige. So waren u.a. Friedrich der Große und Napoleon begeisterte Sammler von Flötenuhren. Auch im Wilhelmsbau ist diese frühe Form der Musikautomaten vertreten. Das Bild auf der gegenüberliegenden Seite zeigt einen Blick in das Uhrenkabinett, in dem zahlreiche wertvolle Originale ausgestellt sind.

Flute-playing Instruments

The first devices that, in a wide sense, can be considered automatic musical instruments were the musical movements built into clocks. Initially, they were chimes or psaltery-related harp clocks. But because of their limited range of sound these instruments could only play very simple melodies. In the 18th century, serinettes, also referred to as "bird organs", became quite popular. Amongst other things they were used to teach canaries to sing. An example from the Wilhelmsbau is shown in the picture at the bottom left.

Serinettes are already equipped with all parts which make up an automatic musical instrument: a drive (hand crank), a sound carrier (pinned barrel), and sound sources (pipes). The hand crank drives both the barrel and the pump. The pinned barrel is sensed by small levers which are attached to valves. If a lever is raised by a pin, the valve opens and compressed air streams into the corresponding pipe.

Refinement of this principle finally led to the development of flute-playing clocks which were driven by weights and had a clockwork with additional flute pipes. Reproduction of melodies was more and more perfected by improving the pinned barrels and the sensing mechanism. Even famous composers like Mozart, Haydn, Händel or Beethoven wrote music especially for flute-playing clocks. The most valuable masterpieces could be found in the castles of emperors and kings. Frederick the Great and Napoleon both were enthusiastic collectors of flute-playing clocks. This early form of automatic musical instruments is also represented in the Wilhelmsbau. The picture on the opposite page shows the "Clock Cabinet" where many valuable originals are on display.

Gegenüberliegende Seite: Das Uhrenkabinett im Wilhelmsbau. Bild links: Eine Serinette oder „Vogelorgel" aus dem 18. Jahrhundert.

Opposite page: The "Clock Cabinet" in the Wilhelmsbau. Picture left: A serinette or "bird organ" from the 18th century.

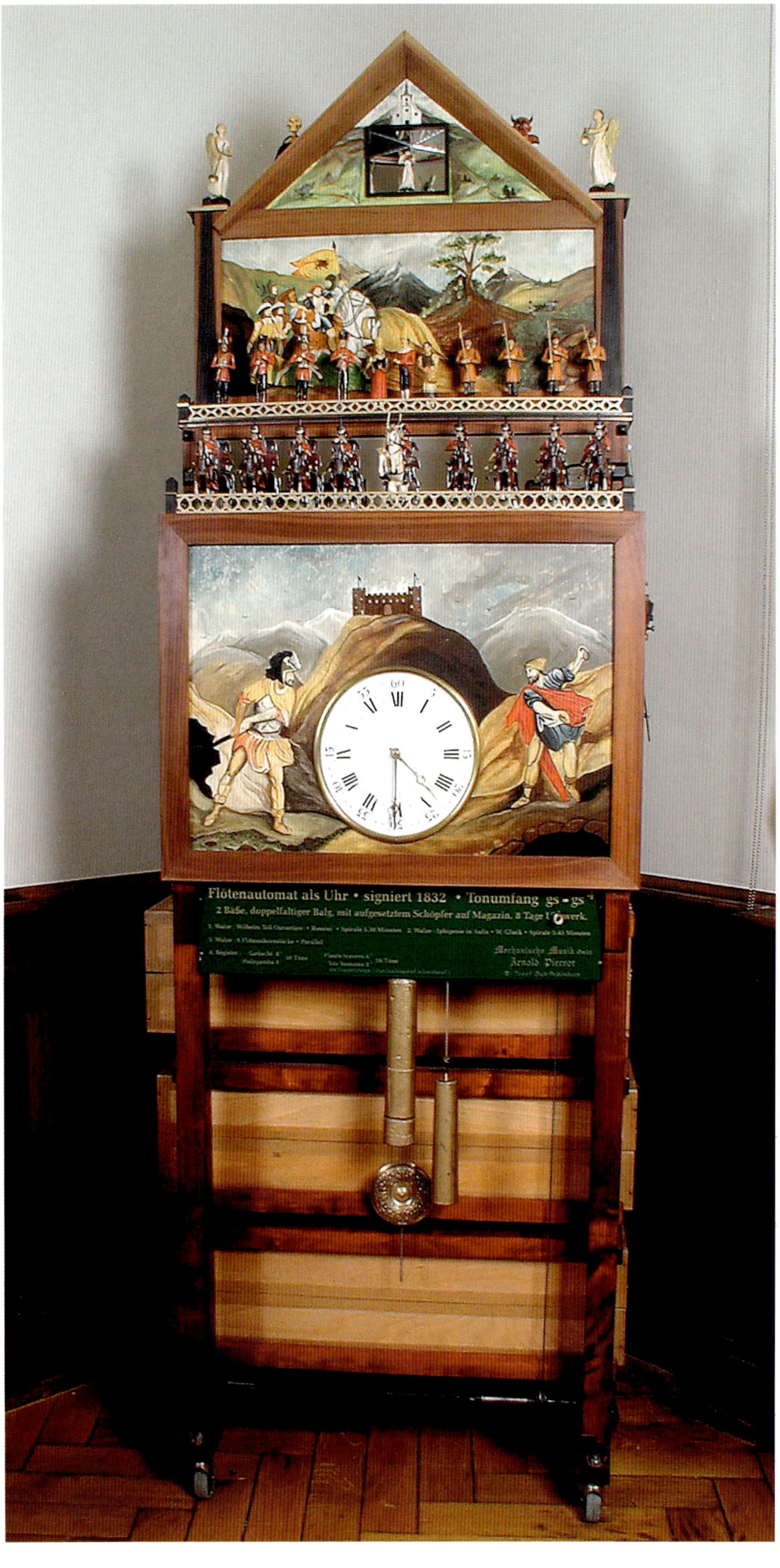

Flötenautomat mit Uhr
Flute-playing Instrument with Clockwork

- Flötenautomat mit Holzpfeifen und Uhrwerk gebaut um 1832. Renovation 1999 durch die Fa. Arnold.
- Maße (HxBxT): 247 cm x 79 cm x 60 cm
- Komplizierter Mechanismus zum Abspielen von spiralförmig bestifteten Walzen. In den Holzkästen unterhalb des Geräts befinden sich drei Ersatzwalzen.
- Das Gerät spielt die „Wilhelm-Tell-Ouvertüre" von Rossini, „Iphigenie auf Tauris" von Willibald Gluck und die Arie "In diesen heiligen Hallen" aus der Zauberflöte von Mozart

- Flute-playing automatic instrument with wooden pipes and clockwork built around 1832. Renovated by Messrs. Arnold in 1999.
- Dimensions (HxWxD): 247 cm x 79 cm x 60 cm
- Sophisticated mechanism for playing spiral pinned barrels. In the wooden cases below the instrument there are three spare barrels.
- The instrument plays the "Overture to Wilhelm Tell" composed by Rossini, "Iphigenia on Tauris" by the composer Willibald Gluck, and "In these August Halls" from the Magic Flute by Mozart

Oben: Die wesentlichen Elemente einer Flötenuhr. (1) Seilzug für das Antriebsgewicht, (2) Drehzahlregulierung, (3) Walzenantrieb, (4) Stiftwalze, (5) Schöpfwerk, (6) Abtasthebel, (7) Holzpfeifen, (8) Uhrwerk.

Top: The essential parts of a flute-playing clock. (1) Rope with attached weight which drives the instrument, (2) speed control, (3) barrel drive, (4) pinned barrel, (5) pump, (6) levers of the sensing mechanism, (7) wooden pipes, (8) clockwork.

- Flötenuhr, gebaut 1854 von Johann Schlegel (1794 - 1868) aus Neustadt im Schwarzwald. Die Figurengruppe mit zwei tanzenden Paaren, vier Musikanten und einem Dirigenten dreht sich zum Takt der außergewöhnlich wohlklingenden Musik.
- Maße (HxBxT): 89 cm x 51 cm x 38 cm

- Flute-playing clock built in 1854 by Johann Schlegel (1794 - 1868) of Neustadt in the Black Forest. The group of figures, consisting of two dancing couples, four musicians and a conductor, is turning around and around in time to the exceptionally well-sounding music.
- Dimensions (HxWxD): 89 cm x 51 cm x 38 cm

- Flötenuhr mit bewegten Figuren, gebaut 1847 ebenfalls von Johann Schlegel. Der Dirigent öffnet und schließt den Mund, zwei Aphroditen bewegen die Arme, zwei Negerinnen klatschen den Takt zu dem vier Musikanten, zwei Soldaten und zwei Posaunenengel spielen.
- Maße (HxBxT): 85 cm x 45 cm x 30 cm

- Flute-playing clock with animated figures built in 1847 also by Johann Schlegel. The conductor opens and closes his mouth, two aphrodites move their arms, two female negroes clap the beat to which four musicians, two soldiers, and two trombone angels are playing.
- Dimensions (HxWxD): 85 cm x 45 cm x 30 cm

- Flötenuhr mit bewegten Figuren, gebaut 1872 von Franz Böhler (1816-1879) aus Neustadt im Schwarzwald. Einschubwerk mit Viertelstundenschlag, das von zwei Mönchen imitiert wird. Klatschender Harlekin, Kuckuck, sechs bewegte Musikanten und zwei Posaunenengel, die im Takt die Fanfaren heben. Auf und absteigender Tod und Teufel. Große Walze mit acht Liedern. Maße (HxBxT): 100 cm x 60 cm x 40 cm.

- Flute-playing clock with animated figures built 1872 by Franz Böhler (1816-1879) of Neustadt in the Black Forest. Slide-in clockwork chiming every quarter imitated by two monks. Clapping harlequin, cuckoo, six animated musicians, and two trombone angels that raise their fanfares to the beat. Raising and descending Death and Devil. Large barrel with eight tunes. Dimensions (HxWxD): 100 cm x 60 cm x 40 cm.

- Flötenuhr mit bewegten Figuren gebaut um 1830. Das Zifferblatt repräsentiert die vier Jahreszeiten. Die Musik, in der sich die Jahreszeiten ebenfalls widerspiegeln, wird durch die Bewegung der sechs animierten Musikanten und das Klatschen des Harlekins versinnbildlicht.
- Maße (HxBxT): 88 cm x 62 cm x 30 cm

- Flute-playing clock with animated figurines built around 1830. The face symbolizes the four seasons. The music, which is also reflecting the seasons, is symbolized by the movement of the six animated musicians and the clapping of the harlequin.
- Dimensions (HxWxD): 88 cm x 62 cm x 30 cm

- Astronomische Flötenuhr mit dem Thema „Die drey Grazien". Sie zeigt den Tag, den Monat, die Mondphase und das Tierkreiszeichen. Die Figurengruppen links und rechts, die die Sonne und den Mond tragen, drehen sich während des Spiels.
- Maße (HxBxT): 95 cm x 58 cm x 32 cm

- Astronomical flute-playing clock with the theme "The Three Graces". It shows the day, the calendar month, the lunar phase and the sign of the zodiac. The groups of figures on the left and right, carrying the sun and the moon, revolve on their axis while the instrument is playing.
- Dimensions (HxWxD): 95 cm x 58 cm x 32 cm

- Besonders seltene und kostbare Flötenuhr von C. E. Kleemeyer, ehemals Hofuhrmacher der preußischen Könige Friedrich II. und Friedrich Wilhelm II. Ein herausragendes Beispiel der Uhrenbaukunst des 18. Jahrhunderts.
- Maße (HxBxT): 55 cm x 41 cm x 39 cm

- Extremely rare and valuable flute-playing clock made by C. E. Kleemeyer, clockmaker and purveyor to the royal household of the Prussian kings Frederick II. and Frederick Wilhelm II. An outstanding example of the art of clock making of the 18th century.
- Dimensions (HxWxD): 55 cm x 41 cm x 39 cm

Wiener Flötenschrank
Flute-playing Instrument from Vienna

- Gebaut um 1810 in Wien
- Das Gehäuse aus massivem Eichenholz mit historischer Farbverglasung stammt vermutlich aus der Zeit um 1860
- Maße (HxBxT): 236 cm x 125 cm x 75 cm
- Walzenbreite 56,5 cm, Walzendurchmesser 13,5 cm, drei Ersatzwalzen vorhanden
- Flötenwerke wurden auch als eigenständige Klangerzeuger ohne Uhrwerk in Musikautomaten verwendet. Ein wunderbares Beispiel ist dieser tadellos erhaltene Flötenautomat vom Anfang des 19. Jahrhunderts.

- Built around 1810 in Vienna
- The cabinet made from massive oak wood with historical color glazing was probably built around 1860
- Dimensions (HxWxD): 236 cm x 125 cm x 75 cm
- Barrel width 56.5 cm, barrel diameter 13.5 cm, three spare barrels available
- Flute pipes were also used separately without a clockwork in automatic musical instruments. A beautiful example is this perfectly preserved flute-playing instrument from the early 19th century.

Die Flöten sind aus Holz gefertigt. Das Gewicht treibt sowohl die Stiftwalze als auch das unmittelbar unter der Walze angeordnete Schöpfwerk an. Im Unterteil des Gehäuses befinden sich drei Ersatzwalzen.

The flutes are made from wood. The weight not only drives the pinned barrel but also the pump that is situated right under the barrel. In the lower part of the cabinet there are three spare barrels.

Orchestrien

Im Wilhelmsbau können die Besucher eine faszinierende Sammlung von Orchestrien aller Epochen und Bauarten erleben, von denen weltweit oft nur wenige Exemplare erhalten sind. Orchestrien verfügen über viele unterschiedliche Klangerzeuger und besitzen daher die Klangvielfalt eines kleinen Orchesters. Typisch sind ein Klavierteil, das bei manchen Geräten auch manuell spielbar ist, Schlaginstrumente wie Trommeln, Becken und Xylophon sowie die unterschiedlichsten Pfeifen. Als Tonträger werden Stiftwalzen oder Papierrollen verwendet. Auch die besten Instrumente mit Plattensteuerung erreichen nicht die Klangfülle eines richtigen Orchestrions. Die meisten Instrumente wurden zwischen 1830 und 1930 in Deutschland gefertigt. Bedeutende Hersteller waren u.a. Hupfeld, Popper und Lösche in Leipzig sowie Welte, Weber, Bruder, Ruth, Imhof & Mukle, Heizmann und Blessing im Schwarzwald und Umgebung. Ab 1930 wurden die Musikautomaten, mit Ausnahme der Kirmes- und Tanzorgeln, zunehmend von Plattenspielern, Tonbändern und Radiogeräten verdrängt.

Visitors can experience a fascinating orchestrion collection in the Wilhelmsbau. There are instruments from all eras and designs of which only a few specimens still exist today. Since orchestrions are equipped with many different sound sources their diversity of sound can be compared to a small orchestra. Typical are a piano part which, in many cases, can also be played manually, percussion instruments like drums, cymbals and xylophones, and different types of pipes. Pinned barrels and paper bands are used as sound carriers. Even the best disc music boxes do not reach the sound quality of an orchestrion. Most of the instruments were built in Germany during 1830 and 1930. Important manufacturers were, e.g., Hupfeld, Popper, and Lösche in Leipzig and Welte, Weber, Bruder, Ruth, Imhof & Mukle, Heizmann, and Blessing in the Black Forest and the surrounding area. After 1930 the automatic musical instruments, except for dance and fairground organs, were more and more displaced by record players, tape recorders, and radio sets.

Orchestrienausstellung im Konzertsaal des Wilhelmsbau. Im Rahmen von Konzertveranstaltungen werden die Instrumente der Öffentlichkeit regelmäßig vorgeführt.

Orchestrion exhibition in the concert hall of the Wilhelmsbau. There are regular concert events during which the instruments are presented to the public.

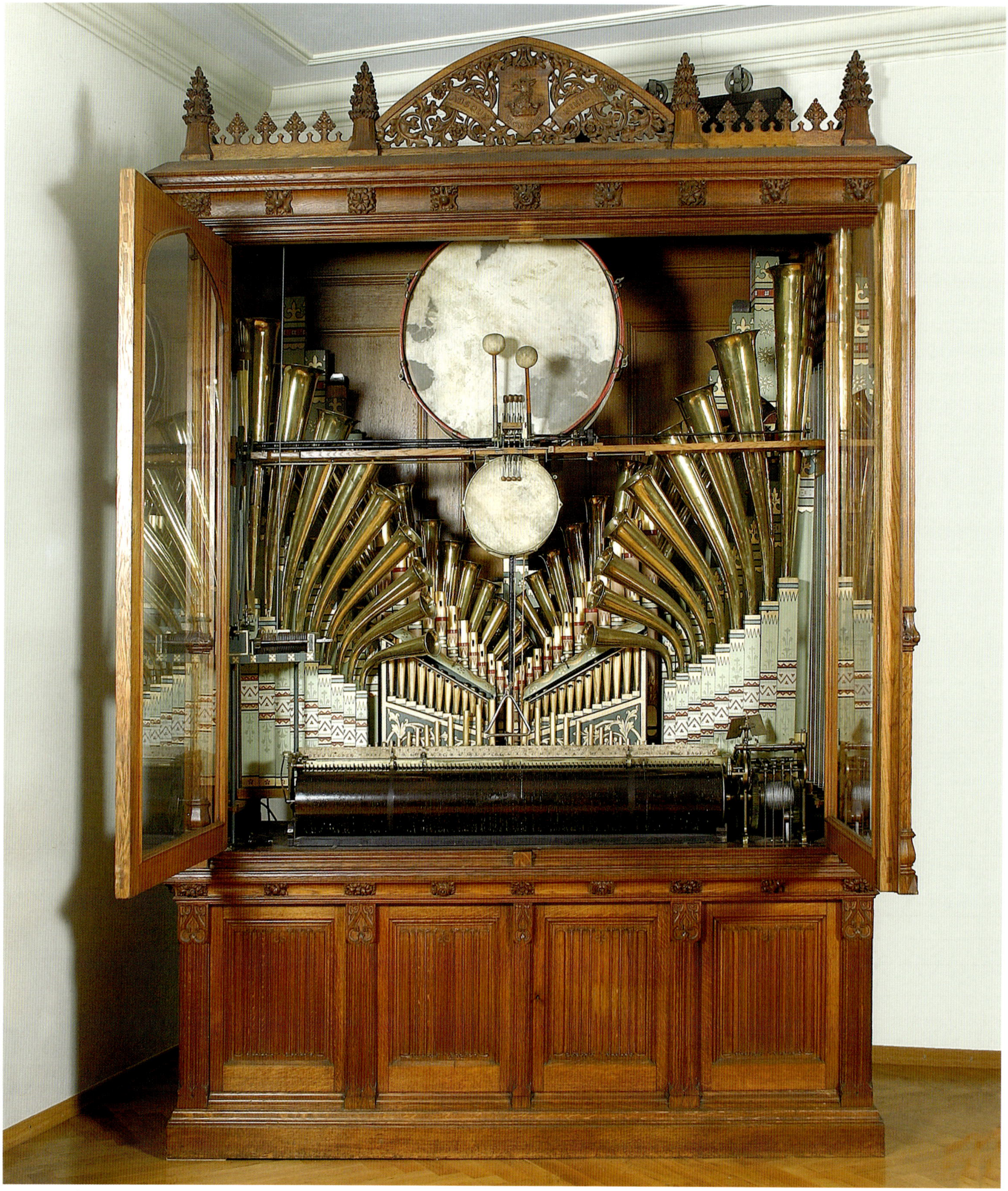

Imhof & Mukle Pfeifenorchestrion
Imhof & Mukle Pipe Orchestrion

- Gebaut 1861 von Imhof & Mukle in Vöhrenbach/Schwarzwald
- Walzengesteuertes Pfeifenorchestrion mit Gewichtsantrieb
- Maße (HxBxT): 360 cm x 240 cm x 160 cm
- Instrumente: 260 Pfeifen mit 98 Tonstufen, drei Trommeln und einer Triangel
- Mit höchstem Aufwand produziertes, edles Saloninstrument aus der Frühzeit der Orchestrien. Bemerkenswert ist die Bemalung der Pfeifen, die bei den meisten Instrumenten später entfernt wurde.
- Der Walzenschrank im gleichen Gehäusestil birgt einen unvorstellbaren, musikhistorischer Schatz: 27 spiralförmig bestiftete Walzen mit Musik aus der Zeit von 1890 bis 1914 mit einer Spielzeit von jeweils ca. sechs bis acht Minuten

- Built 1861 by Imhof & Mukle in Vöhrenbach/Black Forest
- Barrel orchestrion with pipes and weight drive
- Dimensions (HxWxD): 360 cm x 240 cm x 160 cm
- Instruments: 260 pipes with 98 degrees, three drums, one triangle
- Exquisite salon instrument from the early times of orchestrions. The painting of the pipes is a remarkable feature since it was removed from most instruments at a later point in time.
- The barrel closet built in the same design contains an incredible treasure of musical history: 27 spiral barrels with music from the era between 1890 and 1914 with a playing time of 6 - 8 minutes each

Die Pfeifen sind aus Holz und Messing gefertigt. Die Bemalung der Holzpfeifen ist erhalten geblieben. Dies ist außergewöhnlich, da diese bei den meisten Instrumenten zu einem späteren Zeitpunkt entfernt wurde. Neben dem Instrument ist der Schrank mit den Ersatzwalzen aufgestellt, ein ungeheurer musikhistorischer Schatz. Die Walzen haben eine Breite von 120 cm und einen Durchmesser von 20 cm.

The pipes are made of wood and brass. The painting of the wooden pipes has survived. This is remarkable since the painting was removed from most instruments at a later point in time. Next to the instrument is the closet with the spare barrels, an incredible treasure of musical history. The barrels have a length of 120 cm and a diameter of 20 cm.

Imhof & Mukle Pfeifenorchestrion
Imhof & Mukle Pipe Orchestrion

- Gebaut von Imhof & Mukle (Orchestrion No. 2385) um 1892 in Vöhrenbach/ Schwarzwald
- Walzengesteuertes Pfeifenorchestrion mit Gewichtsantrieb
- Gehäuse aus Nussbaum-Wurzelfurnier
- Maße (HxBxT): 310 cm x 180 cm x 100 cm
- Instrumente: 233 Pfeifen und Schlagzeug
- Mit höchstem Aufwand produziertes, edles Saloninstrument
- Walzenbreite 90 cm, Walzendurchmesser 17,5 cm, 17 Ersatzwalzen vorhanden

- Built around 1892 (Orchestrion No. 2385) by Imhof & Mukle in Vöhrenbach/Black Forest
- Barrel orchestrion with pipes and weight drive
- The cabinet is made of walnut root veneer
- Dimensions (HxWxD): 310 cm x 180 cm x 100 cm
- Instruments: 233 pipes and percussion
- Exquisite salon instrument built to the highest standards
- Barrel width 90 cm, Barrel diameter 17.5 cm, 17 spare barrels available

Oben: Für das Instrument sind 17 Originalwalzen vorhanden. Links: Die Walze mit den Abtasthebeln.

Top: There are 17 original barrels available for the instrument. Left: The barrel with the sensing levers.

Gebr. Ellenrieder Tuttlingen

Orchestrion der Gebr. Ellenrieder

- Gebaut von der Firma Gebr. Ellenrieder um 1895 in Tuttlingen
- Walzengesteuertes Orchestrion mit Gewichtsantrieb
- Maße (HxBxT): 270 cm x 160 cm x 90 cm
- Instrumente: Klavier, Mandoline, Xylophon, Triangel, Schlagzeug
- Typisches Gaststätteninstrument mit Münzeinwurf
- Walzenbreite 115 cm, Walzendurchmesser 30 cm, zwei Walzen vorhanden

- Built around 1895 by Ellenrieder Bros. in Tuttlingen
- Barrel orchestrion with weight drive
- Dimensions (HxWxD): 270 cm x 160 cm x 90 cm
- Instruments: Piano, mandoline, xylophone, triangle, percussion
- Typical coin-operated orchestrion for pubs and guesthouses
- Barrel width 115 cm, Barrel diameter 30 cm, two barrels available

Oben: Das Instrument in geschlossenem Zustand. Die Ersatzwalze wird üblicherweise im Unterteil des Schranks aufbewahrt. Links: Der Instrumententeil. Da das Orchestrion keine Pfeifen besitzt, fehlt ein Schöpfwerk und damit auch der entsprechende Antrieb (vergleiche die Antriebseinheit ganz links neben der Walze mit der des Imhof & Mukle Pfeifenorchestrions).

Top: The instrument with closed cabinet. The spare barrel is usually stored in the lower part of the cabinet. Left: The instrument part. Since the orchestrion has no pipes, the pump and its drive are missing (compare the drive unit at the very left of the barrel with the drive unit of the Imhof & Mukle pipe orchestrion).

Gegründet
1813.
L. JACOB·KGL.HOFL.
STUTTGART
Telefon:
1546

Orchestrion der Fa. Lösche

- Gebaut um 1912 von Lösche in Leipzig
- Walzengesteuertes Orchestrion mit Gewichtsantrieb, Walzenbreite 86 cm
- Händler war L. Jacob, Königlicher Hoflieferant aus Stuttgart
- Gehäuse aus deutschem Nussbaum
- Maße (HxBxT): 280 cm x 130 cm x 75 cm
- Instrumente: Klavier, Mandoline, Becken, Triangel und Schlagzeug
- Produktionszeit 1895 - 1914
- Typisches Gaststätteninstrument mit Münzeinwurf
- Nach dem Einschalten öffnen sich die Türen automatisch und der Innenraum wird beleuchtet
- Das Instrument wurde viele Jahre vom Gasthof „Zum Stromberg" in Ochsenbach betrieben und wurde von dort erworben

- Built around 1912 by Lösche in Leipzig
- Barrel orchestrion with weight drive, barrel width 86 cm
- The dealer was L. Jacob of Stuttgart
- Cabinet made of German walnut
- Dimensions (HxWxD): 280 cm x 130 cm x 75 cm
- Instruments: Piano, mandoline, cymbal, triangle, and drums
- Period of production 1895 - 1914
- Typical coin-operated orchestrion for pubs and guesthouses
- After the instrument is switched on, the doors open and the interior is illuminated
- The instrument was used for many years by the restaurant "Zum Stromberg" in Ochsenbach from where it was purchased

·Eldorado V·
Telefon 4393.
Käser & Musser
Stuttgart.
Königliche
Hoflieferanten

„Eldorado V" Orchestrion

- Gebaut um 1880 von der Sächsischen Orchestrionfabrik F. O. Glass in Leipzig
- Walzengesteuertes Orchestrion mit Gewichtsantrieb, Walzenbreite 45,5 cm
- Maße (HxBxT): 230 cm x 90 cm x 80 cm
- Instrumente: Klavier, Mandoline, Xylophon, großes und kleines Becken, Triangel

- Built around 1880 by the Saxon Orchestrionfabrik F. O. Glass in Leipzig
- Barrel orchestrion with weight drive, barrel width 45,5 cm
- Dimensions (HxWxD): 230 cm x 90 cm x 80 cm
- Instruments: Piano, mandoline, xylophone, small and large cymbal, triangle

Orchestrion von F.O. Glass

- Gebaut von der Sächsischen Orchestrionfabrik F.O. Glass in Leipzig
- Walzengesteuertes Orchestrion mit Federantrieb
- Gehäuse aus massivem Eichenholz mit einer Front aus Facettenspiegeln
- Maße (HxBxT): 260cm x 220cm x 90cm
- Typisches Gaststätteninstrument mit Münzeinwurf
- Instrumente: Klavier, Trommeln, Becken und Triangel

- Built by the Saxon Orchestrionfabrik F.O. Glass in Leipzig
- Barrel orchestrion with spring-wound clockwork
- The cabinet with facet mirrors is made of massive oak wood
- Dimensions (HxWxD): 260cm x 220cm x 90cm
- Typical coin-operated orchestrion for pubs and guesthouses
- Instruments: Piano, drums, cymbals, and triangle

„Jazbandor“ Orchestrion

- Gebaut um 1920 von Blessing in Unterkirnach / Schwarzwald
- Walzengesteuertes Orchestrion mit Federantrieb
- Gehäuse aus Mahagoni
- Maße (HxBxT): 230 cm x 136 cm x 71 cm
- Instrumente: Klavier, Klangröhren und Schlagzeug
- Auf der Stiftwalze sind 10 Musikstücke eingeschlagen

- Built around 1920 by Blessing in Unterkirnach / Black Forest
- Barrel orchestrion with spring-wound clockwork
- Cabinet made of mahagony
- Dimensions (HxWxD): 230 cm x 136 cm x 71 cm
- Instruments: Piano, tubular bells, and percussion
- There are 10 pieces of music programmed on the barrel

Orchestrion von Jebavy Truknov

- Gebaut um 1890 von Jebavy Truknov / Böhmen & Mähren
- Walzengesteuertes Orchestrion mit Gewichtsantrieb, Walzenbreite 85,5 cm, Walzendurchmesser 27 cm
- Maße (HxBxT): 280 cm x 120 cm x 60 cm
- Instrumente: Zither, Triangel und Schlaginstrumente

- Built around 1890 by Jebavy Truknov / Böhmen & Mähren
- Barrel orchestrion with weight drive, barrel width 85.5 cm, barrel diameter 27 cm
- Dimensions (HxWxD): 280 cm x 120 cm x 60 cm
- Instruments: Zither, triangle, and percussion

„Atlantic" Orchestrion

- Gebaut um 1900 von Hupfeld in Leipzig
- Walzengesteuertes Orchestrion mit Gewichtsantrieb
- Maße (HxBxT): 250 cm x 80 cm x 70 cm
- Typisches Gaststätteninstrument mit Münzeinwurf
- Instrumente: Klavier, Mandoline, Xylophon, Becken, Trommel und Triangel
- Walzenbreite 45 cm, eine Ersatzwalze vorhanden

- Built around 1900 by Hupfeld in Leipzig
- Barrel orchestrion with weight drive
- Dimensions (HxWxD): 250 cm x 80 cm x 70 cm
- Typical coin-operated orchestrion for pubs and guesthouses
- Instruments: Piano, mandoline, xylophone, cymbal, drum, and triangle
- Barrel width 45 cm, one spare barrel available

Audio CD Titel 5

„Brabo“ Orchestrion

- Gebaut um 1920 von Weber in Waldkirch / Schwarzwald
- Pneumatiksteuerung mit elektrischem Antrieb, perforierte Papierrolle mit 324 mm Breite als Tonträger
- Maße (HxBxT): 280 cm x 180 cm x 90 cm
- Instrumente: Klavier, Mandoline, Xylophon und Violinflöten
- Die Violinflöten und das Xylophon können auch solo spielen. Der Klavierbass sorgt dann für die Begleitung.
- Das Instrument spielt elegante Tanzmusik der 1920er Jahre
- Auf dem Dach sind hölzerne und an der Front gläserne Schwellerjalousien zur Betonung der lauten und leisen Passagen angebracht

- Built around 1920 by Weber in Waldkirch / Black Forest
- Pneumatic control with electro motor, perforated paper roll with a width of 324 mm as sound carrier
- Dimensions (HxWxD): 280 cm x 180 cm x 90 cm
- Instruments: Piano, mandoline, xylophone, and violin pipes
- The violin pipes and the xylophone can also play solo accompanied by the piano bass
- The instrument plays elegant dance music of the 1920s
- There are wooden swell shutters on the top and glass swell shutters at the front to better emphasize the loud and quiet parts of the music

Das Weber „Brabo“ Orchestrion ist ein typischer Vertreter der hochentwickelten Musikautomaten mit Pneumatik und Papierrollensteuerung. Im Unterteil des Geräts ist das elektrisch betriebene Schöpfwerk zu sehen. Die Papierrolle wird durch einen Windmotor angetrieben (beachte in den Bildern rechts den Windbalg neben der Papierrolle).

The Weber "Brabo" Orchestrion is a typical representative of the highly developed automatic instruments with pneumatic and paper roll control. In the lower part of the instrument the pump, which is driven by an electro motor, can be seen. The paper roll is driven by a wind motor (note the bellows besides the paper roll in the pictures at the right).

Audio CD Titel 6 + 7

„Luna" Orchestrion

- Gebaut um 1914 von Popper in Leipzig
- Pneumatiksteuerung mit elektrischem Antrieb, perforierte Papierrolle mit 348 mm Breite als Tonträger
- Riesiges Saloninstrument in edelster Ausführung
- Maße (HxBxT): 340 cm x 220 cm x 110 cm
- Händler war die Fa. Gebrüder Winkler in Gera
- Instrumente: Klavier, Mandoline, 140 Orgelpfeifen in fünf Registern, Xylophone aus Holz und Metall, Schlagzeug
- Im beleuchteten Glasfenster sieht man während des Spiels einen See mit Wasserfall und vorbeiziehenden Schwänen
- Das „Luna" Orchestrion war ein Meisterstück der Fa. Popper. Das Exemplar im Wilhelmsbau ist das einzige, das weltweit noch bekannt ist.

- Built around 1914 by Popper in Leipzig
- Pneumatic control with electro motor, perforated paper roll with a width of 348 mm as sound carrier
- Huge salon instrument built to the highest standards
- Dimensions (HxWxD): 340 cm x 220 cm x 110 cm
- The dealer was Winkler Bros. in Gera
- Instruments: Piano, mandoline, five stops with 140 organ pipes, xylophones made of wood and metal, percussion
- While the instrument is playing, a lake with waterfall and swans can be seen in the illuminated window
- The "Luna" orchestrion was a masterpiece of the Popper company. The specimen in the Wilhelmsbau is the only one known worldwide.

Der Vergleich mit der daneben aufgestellten, lebensgroßen Puppe zeigt die gewaltigen Dimensionen dieses exzellent erhaltenen Orchestrions.

The comparison with the life-sized mannequin doll illustrates the enormous size of this perfectly preserved orchestrion.

Gebrüder Winkler
Gera-R.
Popper's

Gegenüberliegende Seite: Das „Luna" ist mit einer wunderschönen Beleuchtung ausgestattet. Links: Die große Trommel ist nur ein Element des Schlaginstrumententeils. Unten: Das riesige Schöpfwerk wird von einem Elektromotor angetrieben. Im Hintergrund ist ein Teil des Rahmens mit den Klaviersaiten zu sehen.

Opposite page: The "Luna" is equipped with a beautiful illumination. Left: The large drum is only one part of the percussion instruments. Below: The huge pump is driven by an electro motor. In the back a part of the frame with the piano strings is visible.

„Erato“ Orchestrion

- Gebaut um 1907 von Weber in Waldkirch / Schwarzwald
- Pneumatiksteuerung mit elektrischem Antrieb, perforierte Papierrolle mit 274 mm Breite als Tonträger
- Maße (HxBxT): 290 cm x 160 cm x 90 cm
- Instrumente: Klavier, Mandoline, Xylophon und Schlagzeug
- Typisches Unterhaltungsgerät, das in Gaststätten und Tanzcafés aufgestellt wurde

- Built around 1907 by Weber in Waldkirch / Black Forest
- Pneumatic control with electro motor, perforated paper roll with a width of 274 mm as sound carrier
- Dimensions (HxWxD): 290 cm x 160 cm x 90 cm
- Instruments: Piano, mandoline, xylophone, and percussion
- Typical commercial orchestrion that was placed in pubs, guesthouses and dance cafes

„Bianca" Orchestrion

- Gebaut um 1917 von Popper in Leipzig
- Pneumatiksteuerung mit elektrischem Antrieb, perforierte Papierrolle mit 345 mm Breite als Tonträger
- Maße (HxBxT): 240 cm x 140 cm x 90 cm
- Instrumente: Klavier, Mandoline, Xylophon, Glockenspiel und Schlagzeug
- Vom „Bianca" gibt es viele unterschiedliche Gehäusevariante. Es wurde üblicherweise in Gaststätten und Tanzcafés aufgestellt.

- Built around 1917 by Popper in Leipzig
- Pneumatic control with electro motor, perforated paper roll with a width of 345 mm as sound carrier
- Dimensions (HxWxD): 240 cm x 140 cm x 90 cm
- Instruments: Piano, mandoline, xylophone, chimes, and percussion
- The "Bianca" was built in many different cabinet styles. It was usually found in pubs, guesthouses and dance cafes.

„Roland" Orchestrion

- Gebaut um 1928 von Popper in Leipzig
- Pneumatiksteuerung mit elektrischem Antrieb, perforierte Papierrolle mit 351 mm Breite als Tonträger
- Maße (HxBxT): 290 cm x 300 cm x 100 cm
- Instrumente: Mit Betonung spielendes Klavier, Mandolineneffekt, Violinflöten und Schlagzeug
- Als Besonderheit verfügt das Gerät über eine Lotosflöte. Hierbei handelt es sich um eine spezielle Kolbenpfeife, bei der die Tonhöhe mit Vibrato permanent variiert wird. Hierdurch entsteht der Eindruck einer menschlichen Begleitstimme.
- Auf dem Dach sind hölzerne und an der Front gläserne Schwellerjalousien zur Betonung der lauten und leisen Passagen angebracht
- Das „Roland" stellt technisch und musikalisch die Krone der mechanischen Musikinstrumente dar. Es entstand auf dem technischen Höhepunkt des Orchestrionbaus, der kurze Zeit später während der Weltwirtschaftskrise ein jähes Ende fand.

- Built around 1928 by Popper in Leipzig
- Pneumatic control with electro motor, perforated paper roll with a width of 351 mm as sound carrier
- Dimensions (HxWxD): 290 cm x 300 cm x 100 cm
- Instruments: Reproducing piano, mandoline, violin pipes, and percussion
- One special feature of this instrument is a so-called lotus flute, a stopped flute equipped with a tremolo device that imitates a human supporting voice by a haunting, rising, and falling tone
- There are wooden swell shutters on the top and glass swell shutters at the front to better emphasize the loud and quiet parts of the music
- With respect to both technical and musical features the "Roland" marked the high spot of automatic musical instruments. It was built at the technical climax of orchestrion manufacturing which shortly after came to a sudden end during the Great Depression.

Oben: Das Instrument wird beim Spiel prachtvoll beleuchtet. Links: Die gläserne Schwellerjalousie an der Frontseite. Daneben: Mit der Lotosflöte kann eine menschliche Begleitstimme imitiert werden.

Top: The instrument is magnificently illuminated while it is playing. Left: The glass swell shutters at the front. Next to it: With the lotus flute a human supporting voice can be imitated.

„Superba“ Orchestrion

- Gebaut um 1920 von Popper in Leipzig
- Pneumatiksteuerung mit elektrischem Antrieb, perforierte Papierrolle mit 350 mm Breite als Tonträger
- Maße (HxBxT): 260 cm x 180 cm x 100 cm
- Instrumente: Klavier, Mandoline, großes Holzxylophon, kleines Metallxylophon

- Built around 1920 by Popper in Leipzig
- Pneumatic control with electro motor, perforated paper roll with a width of 350 mm as sound carrier
- Dimensions (HxWxD): 260 cm x 180 cm x 100 cm
- Instruments: Piano, mandoline, large wooden xylophone, small metal xylophone

„Unika" Orchestrion

- Gebaut um 1920 von Weber in Waldkirch / Schwarzwald
- Pneumatiksteuerung mit elektrischem Antrieb, perforierte Papierrolle mit 274 mm Breite als Tonträger
- Maße (HxBxT): 280 cm x 170 cm x 90 cm
- Instrumente: Klavier, Mandolineneffekt und Violinenpfeifen
- Populäres Instrument, das insbesondere in Cafés, Restaurants und Tanzhallen aufgestellt wurde
- Der Aufsatz gehört in den Mittelteil des Geräts. Er wurde nach oben gesetzt, um einen Blick auf den Instrumententeil zu ermöglichen.

- Built around 1920 by Weber in Waldkirch / Black Forest
- Pneumatic control with electro motor, perforated paper roll with a width of 274 mm as sound carrier
- Dimensions (HxWxD): 280 cm x 170 cm x 90 cm
- Instruments: Piano, mandoline effect, and violin pipes
- A very popular instrument that was mostly placed in cafes, restaurants, and dance halls
- The top belongs into the middle part of the orchestrion. It was placed on top to allow a better view of the instrument section.

The happy Jazz-band

„Konzertist" Orchestrion

- Gebaut um 1913 von der Firma Popper in Leipzig
- Pneumatiksteuerung mit elektrischem Antrieb, perforierte Papierrolle mit 350 mm Breite als Tonträger
- Maße (HxBxT): 245 cm x 160 cm x 95 cm
- Ausführung „Happy Jazz" mit Mahagonigehäuse und Schlagzeugaufsatz
- Instrumente: Klavier, Mandoline und Schlagzeug
- Rollen mit mehr als 1000 Musikstücken vorhanden

- Built around 1913 by Popper in Leipzig
- Pneumatic control with electro motor, perforated paper roll with a width of 350 mm as sound carrier
- Dimensions (HxWxD): 245 cm x 160 cm x 95 cm
- "Happy Jazz" version with percussion top and a cabinet made of mahagony
- Instruments: Piano, mandoline, percussion
- Music rolls with more than 1000 pieces of music available

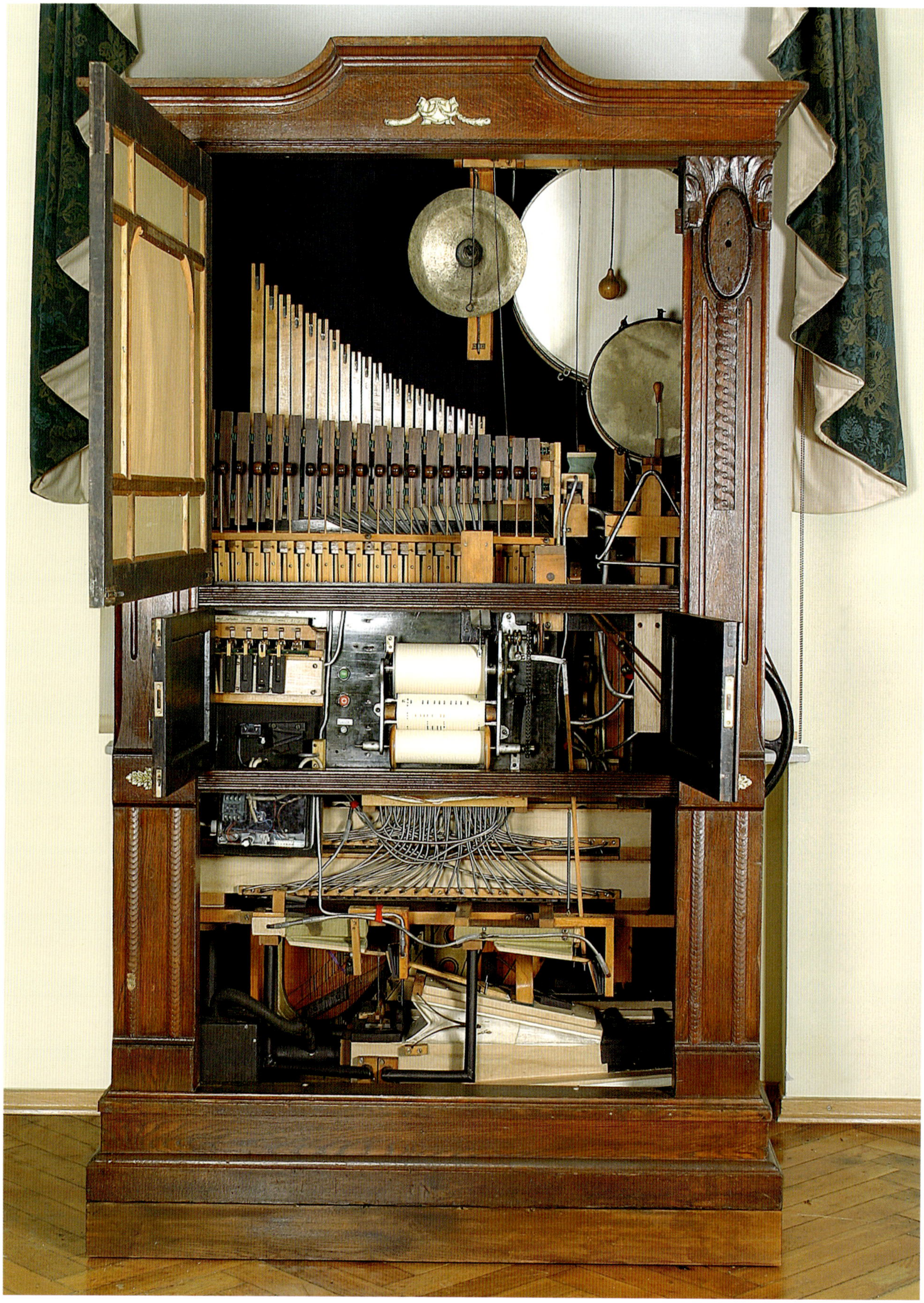

„Pepita“ Orchestrion

- Gebaut um 1908 von Hupfeld in Leipzig
- Pneumatiksteuerung mit elektrischem Antrieb, perforierte Papierrolle mit 200 mm Breite als Tonträger
- Maße (HxBxT): 270 cm x 150 cm x 80 cm
- Instrumente: Klavier, Xylophon, Triangel, Trommel und Violinflöten
- Das Instrument spielt wilhelminisch-preußische Musik aus der Zeit des letzten deutschen Kaisers Wilhelm II.

- Built around 1908 by Hupfeld in Leipzig
- Pneumatic control with electro motor, perforated paper roll with a width of 200 mm as sound carrier
- Dimensions (HxWxD): 270 cm x 150 cm x 80 cm
- "Happy Jazz" version with percussion top and a cabinet made of mahagony
- Instruments: Piano, xylophone, triangle, drums, and violin pipes
- The instrument plays Wilhelminian-Prussian music from the time of the last German emperor Wilhelm II.

Oben: Auch das „Pepita“ Orchestrion wurde mit einer Beleuchtung versehen. Rechts: Das Oberteil des Instruments ist mit Facettenspiegeln und einer dekorativen Druckgrafik geschmückt.

Top: The "Pepita" orchestrion was also equipped with illumination. Right: The top of the instrument is decorated with facet mirrors and a printed illustration.

Audio CD Titel 10

„Commandant“ Orchestrion

- Gebaut um 1930 von Imhof & Mukle in Vöhrenbach / Schwarzwald
- Pneumatiksteuerung mit elektrischem Antrieb, perforierte Papierrolle mit 285 mm Breite als Tonträger
- Maße (HxBxT): 200 cm x 160 cm x 90 cm
- Instrumente: Klavier, Mandolineneffekt und Violinenpfeifen
- Zusätzlich ausgestattet mit Schwellerjalousien und automatischer Rollenführung
- Das Gerät wurde unvollständig aus der Konkursmasse des Herstellers erworben und durch die Firma Arnold fertig gestellt

- Built around 1930 by Imhof & Mukle in Vöhrenbach / Black Forest
- Pneumatic control with electro motor, perforated paper roll with a width of 285 mm as sound carrier
- Dimensions (HxWxD): 200 cm x 160 cm x 90 cm
- Instruments: Piano, mandoline effect, and violin pipes
- Additionally equipped with swell shutters and an automatic guide for the music rolls
- The incomplete instrument was purchased from the assets of the bankrupt manufacturer and completed by Messrs. Arnold

„Grandezza" Orchestrion

- Gebaut um 1923 von Weber in Waldkirch / Schwarzwald
- Pneumatiksteuerung mit elektrischem Antrieb, perforierte Papierrolle mit 324 mm Breite als Tonträger
- Maße (HxBxT): 220 cm x 170 cm x 90 cm
- Instrumente: Klavier, Mandolineneffekt, mit Betonung spielendes Xylophon
- Die Musikrollen waren so ausgelegt, dass das Xylophon auch solo spielen konnte. Manche der Xylophon-Soli sind so komplex, dass sie ein Mensch nicht spielen könnte.

- Built around 1923 by Weber in Waldkirch / Black Forest
- Pneumatic control with electro motor, perforated paper roll with a width of 324 mm as sound carrier
- Dimensions (HxWxD): 220 cm x 170 cm x 90 cm
- Instruments: Piano, mandoline effect, xylophone with expression
- The music rolls were designed to allow the xylophone to be played separately. Some of the xylophone solos are so complex that they cannot be played by a human being.

HUPFELD

Zwei Instrumente haben in der Unterhaltungsmusik des späten 19. und frühen 20. Jahrhunderts eine besondere Rolle gespielt: Die Geige und das Klavier. Entsprechend groß war das Bemühen der Musikautomatenbauer, diese beiden Instrumente in möglichst perfekter Form in ihre Geräte zu integrieren. Die beiden nächsten Kapitel sind daher speziell Musikautomaten gewidmet, bei denen das Klavier- bzw. das Geigenspiel im Vordergrund stand. Den Anfang machen die automatisch spielenden Geigen.

Das größte Problem bei der mechanischen Reproduktion des Geigenklangs sind insbesondere die vielfältigen Nuancen, die ein menschlicher Geigenspieler durch die Variation von Anpressdruck und Geschwindigkeit des Geigenbogens erreichen kann. Die meisten Musikautomatenbauer haben sich daher damit begnügt, Violinen durch spezielle Pfeifen (Violinpfeifen) zu imitieren. Bei der Beschreibung der Instrumentierung wurde bei den vorstehend beschriebenen Orchestrien darauf hingewiesen, wenn solche Pfeifen enthalten sind. Auch die unten gezeigte Pfeifengeige von Hupfeld bedient sich eines solchen Violineneffekts.

Hupfeld hat sich damit aber nicht zufrieden gegeben. Nach jahrelangem Tüfteln gelang es schließlich, Instrumente zu bauen, die echte Geigen perfekt spielen konnten. Es waren diese Instrumente, als Hupfeld Phonoliszt-Violina bezeichnet, die die Firma Hupfeld zur Legende werden ließen. Auf den nächsten Seiten werden die Hupfeld-Geigen, die im Wilhelmsbau zu sehen sind, im Detail beschrieben.

Two instruments were of special importance for playing light music during the late 19th and the early 20th century: the piano and the violin. Therefore, the manufacturers of automatic musical instruments tried hard to incorporate these instruments into their orchestrions in a most perfect way. For this reason, the next two chapters are specially devoted to automatic musical instruments that emphasize the reproduction of piano and violin music. The self-playing violins will make the start.

The biggest problem with the reproduction of the violin sound are the many tone nuances a human player can generate by varying the pressure and the speed of the violin bow. Therefore, many manufacturers of automatic instruments contented themselves by imitating the violin sound with special pipes (violin pipes). In the description of the aforementioned instruments it was indicated whether they contained special violin pipes. The Hupfeld pipe violin shown below also creates a violin effect by using violin pipes.

Hupfeld, however, did not stop here. After many years of hard efforts they finally succeeded in building automatic musical instruments that could play real violins in a perfect way. It were these very instruments, called Hupfeld Phonoliszt-Violina, that turned the Hupfeld company into a legend. The next pages will describe the Hupfeld violins that are on display in the Wilhelmsbau in detail.

Pfeifengeige - Pipe Violin

- Gebaut um 1912 von Hupfeld in Leipzig
- Pneumatiksteuerung mit elektrischem Antrieb, perforierte Papierrolle als Tonträger
- Maße (HxBxT): 260 cm x 180 cm x 100 cm
- Die Instrumentierung besteht aus einem Klavier und einem Geigeneffekt, bei dem der Geigenton durch spezielle Pfeifen nachgeahmt wird. Interessanterweise wurde die Pfeifenvioline von Hupfeld parallel zu der Phonoliszt-Violina mit echten Geigen produziert, wobei alle Instrumente die gleichen Rollen mit einer Breite von 295 mm verwenden.
- Extrem seltenes Instrument, weltweit nur drei Exemplare bekannt

- Built around 1912 by Hupfeld in Leipzig
- Pneumatic control with electro motor, perforated paper roll as sound carrier
- Dimensions (HxWxD): 260 cm x 180 cm x 100 cm
- The Hupfeld pipe violin contains a piano part and a violin effect, which imitates the violin sound with special pipes. Interestingly, Hupfeld built the pipe violin parallel to the Phonoliszt-Violina that contained real violins. All Hupfeld violin instruments use the same music rolls with a width of 295 mm.
- A very rare instrument, only three specimens are known worldwide

HUPFELD
PHONOLISZT-VIOLINA

Selbstspielende Geigen

Im Wilhelmsbau kann eine der weltweit größten Sammlungen selbstspielender echter Geigen bewundert werden. Die Ausstellung umfasst drei Hupfeld Phonoliszt-Violina (Modell A, Modell B und Doppelgeige) und eine Mills Violano-Virtuoso.

Selbstspielende Geigen gehören zu den faszinierendsten Musikautomaten, da ein natürliches Geigenspiel technisch außerordentlich schwer reproduziert werden kann. Der Durchbruch auf diesem Gebiet gelang der Firma Hupfeld mit der Phonoliszt-Violina im Jahr 1908. In diesem Instrument wurde ein Klavier mit drei Geigen kombiniert, bei denen jeweils eine der vier Saiten von einem kreisförmigen, mit 1344 Rosshaarsträhnen bespannten Bogen angestrichen wird. Das Geigenspiel ist so lebensecht, dass die Phonoliszt-Violina selbst von Konzertgeigern als achtes Weltwunder bezeichnet wurde.

Die Hupfeld Geigeninstrumente gehören heute zu den seltensten und kostbarsten Musikautomaten. Von der Phonoliszt-Violina Modell A existieren weltweit nur noch ca. 16 Exemplare. Das Instrument im Wilhelmsbau ist aber nicht nur äußerst selten, sondern gehört auch zu den ganz wenigen Hupfeld Geigen, die noch voll spielfähig sind.

Self-playing Violins

In the Wilhelmsbau, one of the world's largest collections of self-playing real violins can be admired. The exhibition includes three Hupfeld Phonoliszt-Violina (model A, model B, and double violin) and a Mills Violano-Virtuoso.

Self-playing violins are some of the most fascinating automatic musical instruments since the reproduction of a natural violin sound is extremely difficult. The breakthrough in this field was the Phonoliszt-Violina introduced by Hupfeld in 1908, which combined three violins with a piano. On each violin one of the four strings is played by a circular bow holding 1,344 strands of horsehair. The violin play is so lifelike that even professional violin players described the Phonoliszt-Violina as the eighth Wonder of the World.

Nowadays, the Hupfeld violin instruments belong to the rarest and most valuable automatic musical instruments. Of the Phonoliszt-Violina model A there are only about 16 instruments worldwide that survived until today. The specimen in the Wilhelmsbau is not only extremely rare, it also belongs to the very few Hupfeld violins that are still completely functional.

Bei der Hupfeld Phonoliszt-Violina wird bei jeder Geige eine Saite gespielt. Die Tonhöhe wird durch 11 pneumatisch gesteuerte Metallgreifer festgelegt.

On each violin of the Phonoliszt-Violina one string is played. The pitch is determined by 11 pneumatically controlled metal fingers.

HUPFELD
PHONOLISZT-VIOLINA

Hupfeld Phonoliszt-Violina Modell A

- Erste Modellversion mit einem orchestrionähnlichen Gehäuse, gebaut ab 1908
- Maße (HxBxT): 228 cm x 152 cm x 89 cm
- Kombiniert ein Klavier mit drei Geigen bis zur Tonhöhe Gis^3
- Pneumatiksteuerung mit elektrischem Antrieb, perforierte Papierrolle als Tonträger
- Gesamtzahl der gebauten Phonoliszt-Violina Instrumente (Modelle A bis C) von 1908 bis 1930 ca. 3000 Stück, weltweit nur ca. 60 Stück erhalten
- Viele Musikrollen von Klassik bis Unterhaltungsmusik verfügbar

- First model with an orchestrion-like cabinet built from 1908
- Dimensions (HxWxD): 228 cm x 152 cm x 89 cm
- Combines a piano with three violins with a pitch up to G sharp 3
- Pneumatic control with electro motor, perforated paper roll as sound carrier
- Between 1908 and 1930 about 3,000 Phonoliszt-Violina (models A - C) were built. About 60 instruments are still existing.
- Many music rolls playing light and classic music available

HUPFELD
HUPFELD
PHONOLISZT-VIOLINA
RÖNISCH

Hupfeld Phonoliszt-Violina Modell B

- Zweite Modellversion mit einem Gehäuse, bei dem die Geigen besser repräsentiert wurden
- Maße (HxBxT): 260 cm x 215 cm x 110 cm
- Technische Ausführung und Instrumentierung wie beim Modell A

- Second model with a cabinet which presented the violins much better
- Dimensions (HxWxD): 260 cm x 215 cm x 110 cm
- The technical design and the instruments were identical with the model A

Hupfeld Phonoliszt-Violina mit Doppelgeige

- Spezialausführung mit zwei parallel spielenden Geigeneinheiten
- Durch die Verdoppelung der Geigen wird ein wesentlich volleres Klangvolumen erreicht
- Konzipiert für größere Restaurants und Cafés
- Maße (HxBxT): 260 cm x 215 cm x 95 cm
- Sonstige technische Ausführung und Instrumentierung wie bei den anderen Hupfeld-Geigen

- Special design with two violin units playing in parallel
- By doubling the number of violins a much fuller sound is achieved
- Intended for use in larger restaurants and cafes
- Dimensions (HxWxD): 260 cm x 215 cm x 95 cm
- Other design features and instruments are identical with the other models

Der zweifache Olympiasieger Markus Wasmeier mit der Hupfeld Doppelgeige bei einem Besuch im Wilhelmsbau.

The double Olympic Champion Markus Wasmeier with the Hupfeld double violin during a visit at the Wilhelmsbau.

Mills Violano Virtuoso

- Automatisch spielendes Geigeninstrument, gebaut von der Mills Novelty Comp., Chicago, USA
- Maße (HxBxT): 180 cm x 105 cm x 79 cm
- Erschien 1906 als reines Geigeninstrument. 1909 wurde ein Klavierteil hinzugefügt.
- Bis 1930 wurden ca. 4500 Exemplare in unterschiedlichen Ausführungen, u. a. auch mit 2 Geigen, gebaut
- Als Tonträger dient eine perforierte Papierrolle mit einer Breite von 354 mm, die mit einer Drahtbürste abgetastet wird
- Durch die Perforation kommt es zu einem Kontakt zwischen den Bürstendrähten und einer Metallschiene. Hierdurch werden elektromagnetische Schalter ausgelöst, die die Instrumente betätigen.

- Self-playing violin instrument built by the Mills Novelty Comp., Chicago, USA
- Dimensions (HxWxD): 180 cm x 105 cm x 79 cm
- Introduced in 1906 containing a violin only. The piano part was added in 1909.
- Until 1930, about 4,500 specimens were built in different versions, e.g., also with two violins
- Perforated paper roll with a width of 354 mm that is sensed by a metal brush
- Due to the perforation in the paper roll the wires of the metal brush can come into contact with a metal bar. This contact triggers electromagnetic switches which operate the instruments.

Pflege und Instandsetzung

Anders als in den meisten Museen werden die mechanischen Musikinstrumente in Sinsheim und Speyer nicht nur ausgestellt, sondern auch regelmäßig vorgeführt. Dies ist nur möglich, weil die Geräte regelmäßig gewartet und dadurch in einem spielbereiten Zustand gehalten werden. Nur noch ganz wenige Spezialisten sind heutzutage in der Lage, die erforderlichen Arbeiten durchzuführen. Wir sind deshalb froh, dass sich die Fachstätte historischer Musikautomaten von Herrn Gotthard Arnold in Bad Schönborn unserer Geräte angenommen hat und mit ihrer Arbeit dafür sorgt, dass möglichst viele Besucher den Zauber dieser einzigartigen Instrumente live erleben können.

Aber auch bei der Renovation der Instrumente hat das Team von Herrn Arnold Fantastisches geleistet. Viele Instrumente waren beim Erwerb in einem desolaten Zustand und mussten in oft mehrjähriger Arbeit völlig neu aufgebaut werden. Herausragende Beispiele für das Können der Fachstätte sind der Neuaufbau der Aeolin Grand Konzertorgel in Sinsheim und der Welte Philharmonie Orgel in Speyer. Aber auch der Klangzauber vieler kleinerer Instrumente wäre ohne die Arbeit von Herrn Arnold und seinen Mitarbeitern und Mitarbeiterinnen für immer verloren gegangen.

Maintenance and Renovation

Unlike in most other museums, the automatic musical instruments in Sinsheim and Speyer are not only shown on exhibit but also demonstrated in performances on a regular basis. This is only possible as a result of regular care and maintenance which keeps the instruments in a ready-to-play condition. Only very few specialists are still able these days to perform this service. We are happy, therefore, that the workshop of Mr. Gotthard Arnold in Bad Schönborn has adopted our instruments for maintenance. Thanks to their efforts, many visitors can enjoy the magic of these unique instruments as a live experience.

Mr. Arnold and his team also did a magnificent job in restoring. At acquisition, many of the instruments were in a desolate condition and needed complete renovation, which sometimes took several years. Outstanding examples of the workshop's skills are the renovation of the Aeolian Grand concert organ in Sinsheim and the Welte Philharmonie Organ in Speyer. But also many smaller instruments would have lost their magic of sound forever without the effort of Mr. Arnold and his employees.

Selbstspielende Klaviere

Die Papierrollensteuerung und die Verbesserung der Pneumatik erlaubten den Bau immer perfekterer Musikautomaten. Besonders berühmt wurden neben den Hupfeld-Geigen die Reproduktionsklaviere der Freiburger Firma Welte. Gegen Ende des 19. Jahrhunderts gelang es Edwin Welte und seinem Partner Karl Bockisch ein Aufnahmeverfahren zu entwickeln, mit dem alle Feinheiten des Klavierspiels festgehalten werden konnten. Zwar hatte es schon vorher automatische Klaviere gegeben, diese konnten Musikstücke aber nur recht seelenlos wiedergeben. Versuche, die Anschlagdynamik und die feinen Änderungen der Lautstärke nachzuahmen, mit denen ein menschlicher Pianist ein Musikstück mit Leben erfüllt, waren fehlgeschlagen.

Als Aufnahmegerät verwendeten Welte und Bockisch ein spezielles Klavier mit einer Aufzeichnungseinheit, die alle Aktionen des Pianisten in elektrische Signale umwandelte und auf einer Papierrolle aufzeichnete. Diese Mutteraufnahme wurde danach in eine gestanzte Papierrolle umgewandelt und vervielfältigt. Welte gelang es, für seine Aufnahmen viele der bedeutendsten Pianisten und Komponisten seiner Zeit zu gewinnen, unter ihnen Eugen d'Albert, Wilhelm Backaus, Edward Grieg, der junge Vladimir Horowitz, Gustav Mahler, Richard Strauß, Max Reger, Karl Reinecke und zahlreiche andere. Diese Aufnahmen geben einen einzigartigen Einblick in die Klavierinterpretationen des späten 19. und frühen 20. Jahrhunderts. Zur Wiedergabe der Rollen wurden sogenannte „Vorsetzer" oder Reproduktionsklaviere verwendet. Ein Vorsetzer ist ein separater Kasten mit einer Abspieleinheit für die Papierrollen, der vor das Klavier geschoben wird und über Hebel die Tasten und Pedale betätigt. Bei den Reproduktionsklavieren ist die Abspieleinheit integriert. Der Begriff „Reproduktionsklavier" wird nur für selbstspielende Klaviere verwendet, die mit künstlerischem Ausdruck spielen können.

Wie es Welte und Bockisch gelungen war, mit den damaligen Mitteln solch perfekte Aufnahmen zu machen, werden wir vermutlich niemals erfahren. Das Verfahren war ein streng gehütetes Firmengeheimnis, das neben Welte und Bockisch nur drei weiteren Personen bekannt war. Sie alle haben ihr Wissen mit ins Grab genommen.

Self-playing Pianos

Paper roll control and the improvement of pneumatics allowed the construction of automatic musical instruments that became more and more perfect. Besides the Hupfeld violins, the reproducing pianos of Welte in Freiburg became particularly famous. At the end of the 19th century, Edwin Welte and his partner Karl Bockisch developed a recording technique that made it possible to capture all nuances of piano playing. At that time many automatic pianos had already been built. However, these instruments played more or less like soulless machines. All efforts to imitate the subtle changes of volume and the dynamics of touch characteristic for a human piano player had failed.

For recording, Welte and Bockisch used a specially designed piano with a recording unit which recognized all actions of the piano player and converted them into electrical signals which were then printed on a paper roll. This primary recording then was converted into a punched master paper roll for duplication. Welte succeeded in engaging many of the most renowned pianists and composers of his time for his recordings, amongst them Eugen d'Albert, Wilhelm Backaus, Edward Grieg, the young Vladimir Horowitz, Gustav Mahler, Richard Strauß, Max Reger, Karl Reinecke and many others. These recordings give us a unique insight into piano interpretations of the late 19th and early 20th century. For playing of the rolls either so called "Vorsetzer" or reproducing pianos were used. A "Vorsetzer" is a separate playing device which is put in front of a piano and operates the keys and the pedals by means of levers. Reproducing pianos have a built-in playing device. The term "reproducing piano" is reserved for automatic pianos that can play with artistic expression.

How Welte and Bockisch managed to make such perfect recordings with the limited technology of their time remains a mystery. The procedure was a well-kept company secret that, apart from Welte and Bockisch, was known to three other persons only. They all took their knowledge with them when they died.

Der Pianist Eugen d´Albert (Mitte) bei der Einspielung eines Klavierstücks mit dem Welte-Aufnahmeklavier. Links davon Karl Bockisch, rechts Edwin Welte.

The pianist Eugen d´Albert (center) during a recording session with the Welte recording piano. At the left Karl Bockisch, at the right Edwin Welte.

Steinway-Konzertflügel mit „Welte rot"-Vorsetzer

- Steinway-Konzertflügel von 1869 aus Palisanderholz, ein Prachtinstrument von höchster Qualität und Klangfülle. Länge mit Vorsetzer 350 cm, Breite 280 cm.
- Der Vorsetzer mit der pneumatisch gesteuerten Abspieleinheit für „Welte rot"-Papierrollen entspricht dem Welte Modell von 1904. Er wurde vor dem 1. Weltkrieg gebaut und 1962 restauriert.
- Maße des Vorsetzers (HxBxT): 130 cm x 160 cm x 50 cm

Steinway Grand Piano with „Welte rot"-Vorsetzer

- Steinway grand piano from 1869 made of palisander wood, a magnificent instrument of highest quality and sound volume. Length 350 cm including the "Vorsetzer", width 280 cm.
- The "Vorsetzer" with a playing device for "Welte rot"-paper rolls corresponds to the Welte model of 1904. It was built before WW I and renovated in 1962.
- Dimensions of the "Vorsetzer" (HxWxD): 130 cm x 160 cm x 50 cm

Bilder oben: Der Vorsetzer kann praktisch vor jedes Klavier geschoben werden. Die Betätigung der Klaviertasten und der Pedale erfolgt durch Holzhebel. Bild links: Zum Antrieb der Abspieleinheit dient ein Windmotor. Die Welte Technologie erlebte eine Reihe von Entwicklungsstufen, die u.a. an der Farbe der Musikrollen erkennbar sind. Die frühen Rollen waren rot, die späteren grün. Daher kommen Bezeichnungen wie „Welte rot" oder „Welte grün". Der hier gezeigte Vorsetzer spielt die frühen „Welte grün" Musikrollen. Auch heute noch werden Musikrollen für Welte Instrumente neu produziert. Diese sind meist weiß.

Pictures above: In principle, the "Vorsetzer" can be put in front of any piano. The piano keys and the pedals are operated by wooden levers. Picture left: The playing device is driven by an air motor. The Welte technology went through several stages of development which, among other things, can be recognized by the colour of the music rolls. The early rolls were red (rot), the later rolls were green (grün). This is the reason for terms like "Welte rot" or "Welte grün". The "Vorsetzer" shown here plays the early "Welte grün" music rolls. New music rolls for Welte instruments are still produced today. Most of them are white.

BECHSTEIN-WELTE

Bechstein Flügel mit „Welte grün“-Abspieleinheit

- Baujahr 1922
- Maße (HxBxT): 180 cm x 160 cm x 140 cm
- Die elektrisch betriebene Pumpe für die Pneumatik befindet sich in einem separaten Kasten. Ein Vorsetzer mit Abspieleinheit für „Welte rot“-Papierrollen ist zusätzlich vorhanden
- Maße des Vorsetzers (HxBxT): 102 cm x 122 cm x 50 cm

Bechstein Grand Piano with „Welte grün“-Playing Device

- Built in 1922
- Dimensions (HxWxD): 180 cm x 160 cm x 140 cm
- The pump for the pneumatics run by electricity is located in a separate case. A "Vorsetzer" with a "Welte rot"-playing device is also available.
- Dimensions of the "Vorsetzer" (HxWxD): 102 cm x 122 cm x 50 cm

Links: Die integrierte Abspieleinheit für „Welte grün“-Musikrollen. Oben: Mit dem Vorsetzer können auch „Welte rot“-Musikrollen abgespielt werden.

Left: The built-in playing device for "Welte grün"-music rolls. Top: With the "Vorsetzer" "Welte rot"-music rolls can be played also.

Steinway-Welte Piano

- Reproduktionsklavier mit Abspieleinheit für „Welte rot"-Papierrollen mit einer Breite von 330 mm gebaut um 1914
- Maße (HxBxT): 160 cm x 160 cm x 85 cm
- Sowohl die Abspieleinheit als auch das Schöpfwerk für die Pneumatik sind im Instrument integriert
- Die Welte-Klavierautomatik wurde in besonders hochwertige Klaviere eingebaut. Dieses Klavier von Steinway übertrifft in bezug auf Ton, Gewicht und handwerkliche Ausführung alles, was jemals an derartigen Klavieren gebaut wurde.

- Reproducing piano built around 1914 with a playing device for "Welte rot"-paper rolls with a width of 330 mm
- Dimensions (HxWxD): 160 cm x 160 cm x 85 cm
- The playing device and the pump for the pneumatics are both built into the instrument
- Welte playing devices were built into particularly high-quality pianos. This piano made by Steinway surpasses all other pianos of this kind with respect to sound, weight and craftsmanship.

Feurich-Welte Piano

- Reproduktionsklavier von 1907 mit Abspieleinheit für „Welte rot"-Papierrollen mit einer Breite von 330 mm
- Das von Feurich gebaute Klavier kann auch von Hand gespielt werden
- Maße (HxBxT): 160 cm x 160 cm x 85 cm

- Reproducing piano built in 1907 with a playing device for "Welte rot"-paper rolls with a width of 330 mm
- The piano manufactured by Feurich can also be played manually
- Dimensions (HxWxD): 160 cm x 160 cm x 85 cm

Pianola Piano

- Automatisches Klavier, gebaut um 1900 von Hupfeld in Leipzig
- Maße (HxBxT): 131 cm x 143 cm x 76 cm
- Pneumatische Steuerung, perforierte Papierrolle mit einer Breite von 287 mm als Tonträger
- Der Antrieb erfolgt über die Fußpedale. Dadurch konnten die Musikrollen mit einer beliebigen Geschwindigkeit abgespielt werden. Auf den Musikrollen befindet sich eine Markierung, die das korrekte Tempo anzeigt.
- Diese Art von Automatikklaviere wurde insbesondere zum Erlernen des Klavierspiels verwendet

- Automatic piano built around 1900 by Hupfeld in Leipzig
- Dimensions (HxWxD): 131 cm x 143 cm x 76 cm
- Pneumatic control, perforated paper roll with a width of 287 mm as sound carrier
- The roll player and the pump are driven by the foot pedals. This allows to play the music rolls at any speed. There is a mark on the music rolls which shows the correct speed.
- This type of automatic piano was mostly used to learn piano

Friktionsklavier (unten / below)

- Gebaut 1890 von Hupfeld in Leipzig
- Maße (HxBxT): ca. 133 cm x 150 cm x 70 cm
- Antrieb manuell mittels Handkurbel, perforiertes Kartonband mit einer Breite von 25 cm
- Das Hupfeld-Friktionsklavier war eines der ersten mechanischen Klaviere. Die rein mechanische Steuerung des Anschlags ohne künstlerischen Ausdruck hat sich aber nie durchsetzen können.

- Hand-cranked automatic piano built 1890 by Hupfeld in Leipzig
- Dimension (HxWxD): ca. 133 cm x 150 cm x 70 cm
- Perforated cardboard with a width 25 cm
- The Hupfeld Friktionsklavier was one of the first automatic pianos. However, the purely mechanic control without artistic expression proved to be unsuccessful.

Nienola Klavier (oben)
Nienola Player Piano (top)

- Sonderanfertigung gebaut 1920 von Niendorf / Luckenwalde
- Maße (HxBxT): 104 cm x 154 cm x 200 cm
- Die Abspieleinheit stammt von Philipps / Frankfurt a.M. und wurde auch für die Philipps Duca Klaviere verwendet
- Das Instrument spielt Philipps Duca Papierrollen, Breite 285 mm
- Mit den Fußpedalen werden die Abspielgeschwindigkeit und die Nuancierung geregelt. Der Unterdruck für die Pneumatiksteuerung wird in dem neben dem Klavier aufgestellten Gebläse erzeugt.

- Special design built in 1920 by Niendorf / Luckenwalde
- Dimensions (HxWxD): 104 cm x 154 cm x 200 cm
- The roll player was built by Philipps / Frankfurt a.M. and was also used for the Philipps Duca pianos
- The instrument plays Philipps Duca music rolls, width 285 mm
- The foot pedals control the tempo and the expression. The vacuum for the pneumatics is generated by a pump in a separate case placed beside the piano.

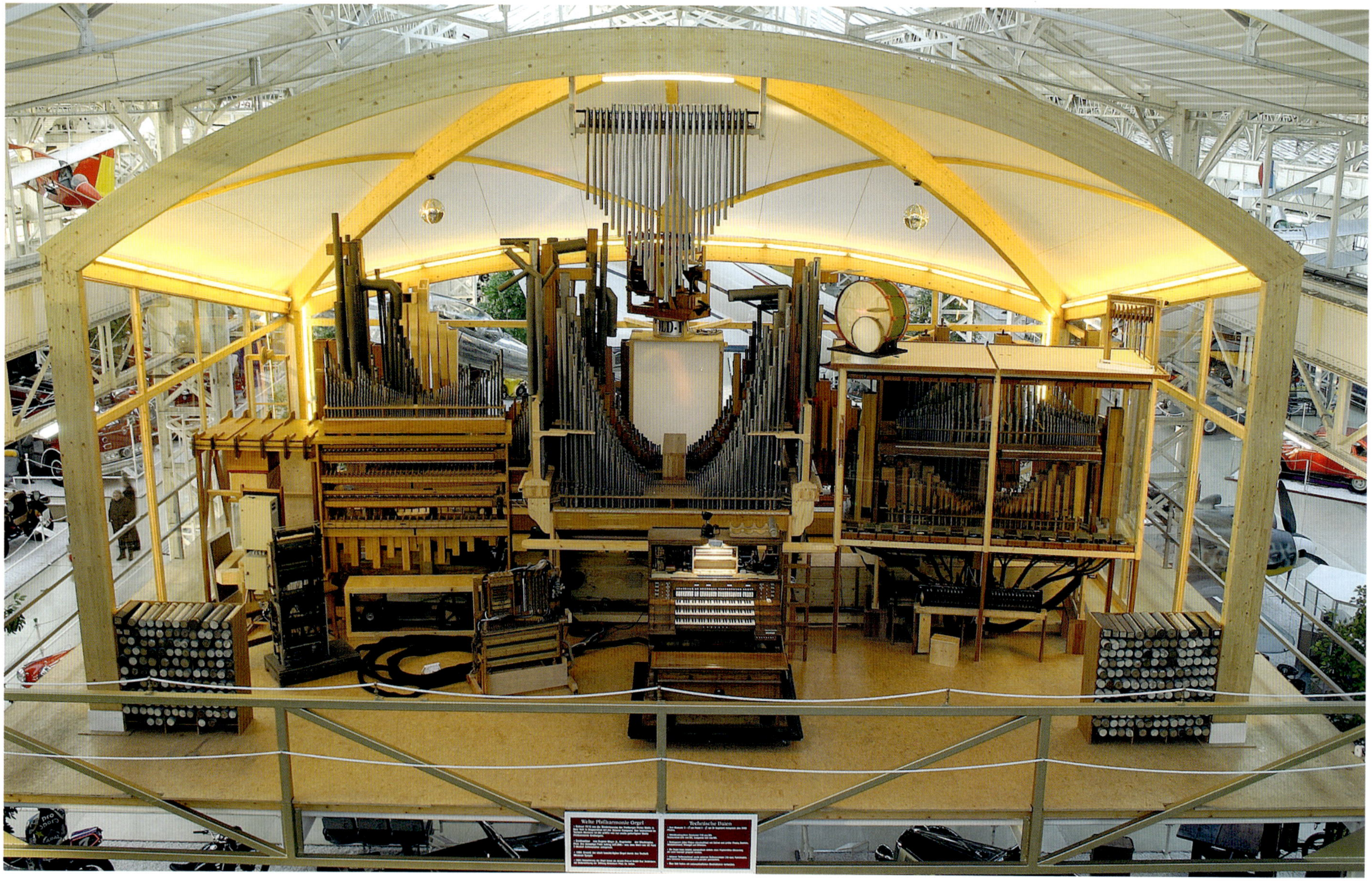

Die Firma Welte hat nicht nur Reproduktionsklaviere und Orchestrien, sondern auch Großorgeln gebaut. Das beeindruckendste Instrument dieser Baureihe war die „Welte Philharmonie", die damals als ein „wahrhaft ideales Hausinstrument für besitzende Kreise" beworben wurde. Aufgrund der gewaltigen Ausmaße war selbst bei herrschaftlichen Häusern oft ein Anbau zum Aufstellen der Orgel erforderlich. Nur wenige Welte Philharmonie Orgeln wurden gebaut, das Museumsstück ist die größte von allen.

Bei der Ankunft in Speyer im Jahr 1993 befand sich das einmalige Instrument mit über 2500 Pfeifen, das sowohl manuell als auch automatisch über eine Papierrollen-Steuerung gespielt werden kann, in einem sehr schlechten Zustand. Zahlreiche wichtige Teile fehlten, die Windkanäle waren verschollen, viele Metallteile wiesen Korrosionen auf und die Verkabelung war weitgehend unbrauchbar geworden. Nach dem Abschluss einer zweijährigen Renovation wurde die jetzt wieder voll funktionsfähige Orgel auf einer großen, absenkbaren Plattform mit Beleuchtung in der Liller Halle aufgestellt.

The Welte company not only built reproducing pianos and orchestrions but also large organs. The most impressive instrument of this kind was the "Welte Philharmonie" which in those days was advertised as "a truly ideal home-music instrument for the wealthy household". Due to its huge dimensions even noble mansions often required an extension to set up the organ. Only a few Welte Philharmonie organs were built, the specimen in the museum is the largest of all.

At its arrival in Speyer in 1993 the unique instrument with more than 2,500 pipes that can be played either automatically with paper-roll control or manually was in very poor condition. Many important parts were missing, the pneumatic system was lost, many metal parts were corroded, and most of the cable connections were no longer usable. After a two years' renovation the organ, which is again fully functional, was placed on a large illuminated platform which can be raised and lowered in the Liller Halle.

Welte Philharmonie Orgel

- Weltweit größte Welte Philharmonie Orgel. Gebaut 1916 von der Welte Niederlassung in New York in Kooperation mit der Skinner Company.
- Maße Orgel (HxBxT): 7 m x 14 m x 10 m. Maße Spieltisch (HxBxT): 1,65 m x 1,45 m x 1,57 m.
- Erstbesitzer war Eugene Meyer jr., Begründer der Washington Post. Der damalige Preis betrug $ 22 000.-, was dem Wert von 50 Ford T-Modell Automobilen entsprach
- 1993 Erwerb der stark beschädigten Orgel durch das Technik Museum Speyer
- 2001/2002 Renovierung durch die Arnold Pierrot GmbH Bad Schönborn mit Unterstützung der Stiftung Rheinland-Pfalz für Kultur
- Drei Manuale C - c^5, ein Pedal C - g^1, 36 Register, 2592 Pfeifen
- Schlagwerk (über Piston einschaltbar) mit kleiner und großer Pauke, Becken, Wirbeltrommel, Triangel und Effekten (drehbar)
- Die Orgel kann sowohl automatisch mittels einer Papierrollen-Steuerung (Rollenbreite 385 mm) als auch manuell gespielt werden
- Interner Rollenautomat sowie externer Rollenwechsler (10-fach Paternoster, zum internen Rollenautomaten parallel geschaltet)
- Über 600 Rollen mit unterschiedlichen Musikstücken vorhanden

Welte Philharmonie Organ

- Largest Welte Philharmonie Organ worldwide. Built 1916 by the Welte subsidiary in New York in cooperation with the Skinner Company.
- Dimensions of the organ (HxWxD): 7 m x 14 m x 10 m. Dimensions of the console (HxWxD): 1.65 m x 1.45 m x 1.57 m.
- The first owner was Eugene Meyer jr., founder of the Washington Post. The price at that time was $ 22,000.-, which was equivalent to the price of 50 Ford Model T automobiles.
- In 1993 the organ was acquired by the Technik Museum Speyer in a very poor condition
- 2001 / 2002 renovation by Messrs. Arnold / Bad Schönborn supported by the Stiftung Rheinland-Pfalz für Kultur
- Three manuals C - c^5, one pedal C - g^1, 36 ranks with 2,592 pipes
- Additional percussion, triangle, and sound effects (rotatable)
- The organ can either be played automatically (paper roll control, roll width 385 mm) or manually
- The player for the paper rolls is built into the console. An external automatic roll changer (10-fold paternoster) can be run in parallel with the built-in roll player.
- More than 600 music rolls available

Oben links: Die Orgelfragmente 1993 nach der Ankunft im Museum. Unten links: Die Renovation der Orgel durch die Firma Arnold benötigte fast zwei Jahre. Oben rechts: Aufbau der Orgel in der Liller Halle im Sommer 2002. Unten rechts: Am 23. Oktober 2002 konnte das jetzt wieder voll spielfähige Instrument einem begeisterten Premierenpublikum präsentiert werden.

Top left: The fragments of the organ after their arrival in the museum in 1993. Bottom left: The renovation of the organ by Messrs. Arnold took almost two years. Top right: Assembly of the Organ in the Liller Halle in summer 2002. Bottom right: On 23. October 2002 the restored instrument could be presented to an enthusiastic first night audience.

432

Urteile von Künstlern

Die bedeutendsten Organisten kamen, um das Instrument zu hören und ihr eigenes Spiel in seinen Musikrollen niederzulegen. In begeisterten Worten äußerten sie ihre Freude über das gelungene Werk.

Prof. Dr. Max Reger, einer der ersten, der die Welte-Philharmonie-Orgel hörte und für sie spielte, schrieb im Juli 1913:

Ich bin ganz entzückt über die Welte-Philharmonie-Orgel und wünsche diesem schönen Instrument die größten Erfolge.

Jos. Bonnet, Paris, schrieb uns im Februar 1913:

En témoignage de vive admiration pour la Merveille qui permettra maintenant d'immortaliser le jeu des virtuoses par une géniale invention, et en toute cordiale sympathie à Messieurs Welte et Bockisch.

PROF. DR. MAX REGER FÜR WELTE-PHILHARMONIE SPIELEND

Prof. Enrico Bossi, Camerlata (Como) schrieb im September 1912:

Ed io, che ho avuto l'occasione di suonare ed udire alcune riproduzioni di importanti composizioni organistiche, posso affermare con piena coscienza che il risultato di tale applicazione è impressionante ed assolutamente superiore ad ogni elogio. La riproduzione dei pezzi eseguiti sull'Organo Welte è tale che ci appaga compiutamente giacchè, non solo l'esecuzione dell'Artista è fissata e riprodotta in modo perfetto, ma il mezzo onde si esplica essendo l'identico usato dall'Artista se risulta che la riproduzione è rigorosamente una ripetizione dell'esecuzione stessa.

Professor Dr. Karl Straube, der bedeutende Organist und Leiter des berühmten Thomaschores in Leipzig, in dieser Stellung einer der Nachfolger von J. S. Bach, schrieb am 14. August 1922:

Dem Mimen flicht die Nachwelt keine Kränze. Seit den Erfindungen des Welte-Mignon-Klaviers und der Welte-Philharmonie-Orgel hat dies Wort für den reproduzierenden musikalischen Künstler seine Gültigkeit verloren. Noch spätere Geschlechter werden die Meister unserer Zeit in ihrem Können, ihrer gesamten Kunstauffassung erkennen lernen können. Mit den Mitteln der Technik sind Vergänglichkeit und Zeit überwunden, ist der Augenblick eines geistigen Erlebens für die Ewigkeit gewonnen worden.

PROF. DR. KARL STRAUBE FÜR WELTE-PHILHARMONIE SPIELEND

Harry Goss Custard, Mus. Bac. Oxon, London, schrieb im Februar 1913:

I feel that I must write a few lines to express my admiration of your wonderful Philharmonic Organ. I consider it, with the Welte-Mignon Piano, the greatest musical invention of modern times. The absolute fidelity with which every variety of touch, every slight shading of tempo, and indeed, everything that goes to the making of an artistic rendering of any work, is reproduced in the record, is truly marvellous. It seems to me that there is only one word which adequately describes your beautiful instruments, and that is "perfection."

Was man vom musikalischen Standpunkt aus über die Welte-Philharmonie-Orgel sagen kann, ist in diesen Äußerungen bedeutendster Orgelkenner und -künstler ausgedrückt.

Urteile der Besitzer von Welte-Orgeln

Die Aufnahme des Instruments und der Künstlerrollen beim Publikum entsprach der des Welte-Mignons. Von den Besitzern kleinerer, mittlerer und größerer Welte-Orgeln liegen eine große Reihe von Urteilen vor, von denen einige hier angefügt seien:

Herr Dr. Karl Höhn, Ulm a. d. D., schreibt unterm 1. November 1921 wie folgt:

Nachdem ich nunmehr meine „Welte-Philharmonie-Orgel Nr. III B" bald 1½ Jahre in täglichem Gebrauch habe, ist es mir ein Herzensbedürfnis, Ihnen mitzuteilen, wie sehr ich mit diesem prachtvollen Instrument zufrieden bin. Sie ist eine ganz grandiose Erfindung, diese Orgel. Ich habe wohl schon ähnliche gesehen und gehört, aber kein Werk, das auch nur entfernt mit Ihrer Philharmonie-Orgel sich messen könnte. Die wunderbare Weichheit des Tones, die Modulationsfähigkeit des Ausdrucks, die Zwischenstufen und der Übergang vom zartesten Pianissimo bis zum mächtigen Fortissimo sind ganz einzigartig.
Schon viele Besucher haben die Orgel gehört, darunter auch Musiker ersten Ranges, und alle waren mit mir einig im Urteil über dieses wohl schönste Musikinstrument, das es gibt. Der Eindruck des Mechanischen ist vollständig verschwunden. Falls Sie es nötig finden, können Sie mich jederzeit als Referenz aufgeben. Einen Fehler nur hat die Orgel, nämlich den, daß sie noch viel zu wenig bekannt ist.

Herr Herm. Theod. Müller, Carioba, schreibt unterm 25. Januar 1924 wie folgt:

Ich habe die Orgel eigenhändig aufgestellt und mit Hilfe meines Bruders zum tadellosen Funktionieren gebracht. Nur zum Stimmen habe ich mir einen Orgelstimmer aus São Paulo kommen lassen.
Ich möchte bei dieser Gelegenheit nicht unterlassen, Ihnen meine Bewunderung für Ihre in jeder Hinsicht hervorragende Arbeit und das staunenerregende Funktionieren des „Welte"apparates auszusprechen. Jeder, dem ich die Orgel vorführte, ist in höchstem Maße entzückt über ihre Wirkung.

Berühmte Besitzer von Welte-Instrumenten - Famous Owners of Welte Instruments

Prinz Bariadinsky

St. Raphael - Cannes

Sir David Salomon

Tunbridge-Wells, England

M. Raefsky
Generalmajor

St. Petersburg, Zarskoje-Selo

Baron de l´Espée

Paris

H.H. The Maharajah of Karpurthala

East India

Ch. Winter

Kopenhagen

Sir Blondel Maple

London

Masch.-Fabrik der K. Ung. Staatseisenbahn

Budapest

Bochumer Gußstahlwerke

Bochum

Hon. J. Mc. Millan

Washington D.C.

Levy Hermanos

Paris - Manila

William Vanderbilt

New York, City

A. S. L. Shields, Esq.

Philadelphia

Atlantic-Garden

New York

Vendome Hotel

Chicago, Illinois

H. H. Maharajah of Patialla

Indien

S. S. Prinz Georg von Griechenland

Kanea

Son Altesse le Khédive d´Egypte

Le Cairo

Rickmers
Reeder

Bremen

Ed. E. Barney

Prest. of Virginia Navigation Co.

Ed. Stocks
Prest. of Hoffmannhouse

New York, City

J. John Grant, Esq.

Cleveland / Ohio

Howard Goud, Esq.

Yacht Niagara

Tito Hékékyan Pacha

Alexandrie - d´Egypte

Harry Miner, Hon.

New York, City

A. Eduard Falz-Fein
Großgrundbesitzer

Südrussland

Claus Spreckles

San Francisco, Cal.

S. M. le Sultan de la Turquie

Konstantinopel

Freiherr von Grote

Villa Candide, Wannsee bei Berlin

Ch. Godderidge
Fabrikant

St. Pölten bei Wien

H. H. the Khedive of Egypt

Kairo

Cornelius Vanderbilt

New Port, U. St. A.

S. H. Sultan von Deli

Sumatra

Sanatorium
Schloss Zimmerberg
Gross-Tabarz i. Th.

Die Frequenz meiner Gäste, welche das von Ihnen gelieferte Orchestrion sehen und bewundern, nimmt mit jedem Tage zu. Ich selbst bin mit dem Instrument, welches ein würdiger Schmuck meines Musiksalons ist, mehr als zufrieden, und spreche Ihnen für die feine und geschmackvolle äussere Ausstattung, sowie die wohlklingende und rhythmisch gut gesetzte Musik meine Anerkennung aus.

Dr. Alex. von Chelstowski.

H. H. the Maharajah
of Patialla, East India.

Howard Goud, Esq.
Yacht Niagara.

Alex. Eduard Falz-Fein
Grossgrundbesitzer, Südrussland.

Freiherr von Grote
Villa Càndide, Wannsee b. Berlin.

etc.

No. 5

Concert-Orchestrion
mit complettem Schlagwerk.

Gehäuse 3,20—3,75 m hoch, 2,10 m breit, 1,12 m tief.

Preis mit 12 Musikrollen . . . . M. 6200.–
Preis der Extra-Musikrollen à . . . „ 52.–

PIANO
5 8285 ⑫
Grubenlichter Waltzes.
Carl Zeller.
Metrostyle.
B 71 73051
Gotterdammerung.
Potpourri.
Arranged by H. Cramer.
Wagner.
The Aeolian Company, New York.
The Orchestrelle Company, London.
Made in U.S.A.
PIANO
5 60905 (D) ⑫
WIENER BLUT.
Walzer, Op. 354.
THE AEOLIAN COMPANY,
Fifth Ave. and 34th St., New York City

Rollenschätze

Ein wichtiges Ziel der Technik Museen in Sinsheim und Speyer ist es, technische Raritäten nicht nur auszustellen, sondern sie auch in einem funktionsfähigen Zustand zu halten. Dies gilt in besonderem Maße für die Sammlung mechanischer Musikinstrumente. Neben einer ständigen Wartung ist hierfür die Verfügbarkeit geeigneter Tonträger eine Grundvoraussetzung, denn ohne Tonträger ist ein Musikautomat so nutzlos wie ein Computer ohne Programme. Über die Jahre ist es dem Museum gelungen, tausende historische Stiftwalzen, Metallplatten und Musikrollen für die Musikautomaten im Museum zusammenzutragen. Das Bild unten zeigt den Rollenraum im Wilhelmsbau, der einen Teil der Sammlung beinhaltet. Dieser einzigartige musikhistorische Schatz gibt einen faszinierenden Einblick in die musikalischen Vorlieben und die künstlerischen Interpretationen längst vergangener Zeiten.

Music Roll Treasures

One important goal of the museums in Sinsheim and Speyer is to show not only technical rarities but also to keep them fully functional. This is particularly true for the collection of automatic musical instruments. Besides continuous maintenance the availability of suitable sound carriers is crucial to achieve this goal. Without a sound carrier an automatic instrument is as useless as a computer without programs. During many years the museum managed to collect thousands of pinned barrels, metal discs and music rolls for its instruments. The picture below shows the music roll room in the Wilhelmsbau which contains part of this collection. This unique treasure of musical history gives a fascinating insight into the musical preferences and artistic interpretations of long past times.

POLYPHON!
SCHUTZ-MARKE. TRADE-MARK
RETREAT AND PRAYER
BORTNIANSKY
4027

Plattenautomaten

Alle frühen Orchestrien wurden durch Stiftwalzen gesteuert, die teuer und schwer zu wechseln waren. Eine flexiblere Alternative erfand Paul Lochmann Ende des 19. Jahrhunderts. Im Jahr 1885 erhielt er ein Patent für Musikautomaten, die eine Lochplatte als Tonträger verwendeten. Solche Platten konnten fabrikmäßig in großen Stückzahlen zu geringen Kosten hergestellt werden. Sie bildeten die Grundlage für die Massenproduktion verhältnismäßig preiswerter Musikautomaten der unterschiedlichsten Größen, sowohl für die gewerbliche Nutzung, z.B. in Gaststätten, als auch für den Heimgebrauch. Die Instrumentierung der Geräte war bewusst einfach gehalten und bestand häufig nur aus Metallkämmen. Sie werden daher auch als Polyphone bezeichnet.

Obwohl die meisten plattengesteuerten Geräte klanglich weniger boten als die Orchestrien und die Schweizer Spieldosen, mit denen sie konkurrierten, setzten sie sich aufgrund des niedrigen Preises und des großen Musikprogramms schnell durch. Das Zentrum der Produktion war Leipzig, wo Lochmann unmittelbar nach der Patenterteilung begann, unter dem Markennamen „Symphonion" plattengesteuerte Musikautomaten zu bauen. Aufgrund des großen Erfolges gründeten zwei seiner Mitarbeiter 1890 die Polyphon Musikwerke, ebenfalls in Leipzig. In den folgenden Jahren entstanden mehrere hunderttausend Geräte, die in die ganze Welt exportiert wurden. Der kometenhafte Aufstieg dieses Industriezweigs endete abrupt nach dem 1. Weltkrieg, als die Grammophone den Markt zu erobern begannen.

Symphonion-Musik-Schrank

No. 154 Sch.

159 Stimmen-Zungen,

ist gegenüber allen Concurrenz-Fabrikaten das denkbar **vollendetste** Spielwerk umfangreicher Art, imponirend durch die **mächtige, packende Klangfülle, meisterhaftes Arrangement** der Musik, **solide** Bauart in gefälligen Formen etc. etc.

Spieldaner jedes Stückes **2** Minuten, so dass dasselbe **vollständig** wiedergegeben wird. **Gesammt**spieldauer nach vollständigem Aufzug der Feder **ca. 45 Minuten.**

Der untere nach vorn aufklappbare Theil des Gehäuses enthält eine Anzahl Fächer zum **Aufbewahren** der Notenscheiben.

Als praktische Neuerung findet der patentirte **Notenscheibenhalter,** welcher an seitlich angcordneten Charnir-Armen drehbar befestigt ist, Verwendung. Das Auflegen der Notenscheibe wird dadurch sehr bequem gemacht.

Note No. 154. (Durchm. 64 cm.)

Symphonion-Musical-case.

No. 154 Sch.

159 tongues.

Glorious tone-effect and richness of sound. The LARGE MUSIC-DISCS permit of all songs, operatic pieces, dances &c. &c. being played with any abbreviation whatsoever. Solid, strong case with store-room for discs.

Caisse à musique Symphonion

No. 154 Sch.

159 lames.

Effet splendide et grande ampleur de sons. Les **grandes feuilles à musique** permettent de jouer sans aucune coupure de nombreux chansons, morceaux d'opéras, danses, etc. Solide et forte caisse à casier pour les feuilles.

	Preise Prix ℳ	Pf
No. 154 Sch. **Symphonion-Musik-Schrank,** 159 Zungen — tongues — lames	440	—
„ 154 Sch. B. **do.** 159 „ Thür mit Butzenscheiben	458	—
„ 192. **do.** 192 Töne, Spieldauer 18 Min. (etwas einf. Gehäuse)	350	—
Musiknoten — tunes — feuilles de musique — zu No. 154 (Durchm. 64 cm)	2	40
do. „ „ „ „ „ „ 192 „ 64 „	2	40

Als Automaten, mit Geldeinwurf, werden obige Schränke ohne Preiserhöhung geliefert.

Price without tunes with case. — **Preis ohne Noten mit Kiste.** — Prix sans feuilles avec caisse.

Bei Baarzahlung innerhalb 30 Tagen 5 %.

5 % discount for cash within 30 days. — Escompte de 5 % pour les payements au comptant dans le délai de 30 jours.

Disc Music Boxes

All early orchestrions used pinned barrels as sound carriers, which were expensive and hard to change. A more flexible alternative was invented by Paul Lochmann at the end of the 19th century. In 1885, he was granted a patent for automatic musical instruments that used perforated discs as sound carriers. Such discs could be produced at low cost in large quantities. They provided the basis for the mass production of relatively cheap automatic musical instruments of different sizes for both commercial use, e.g., in pubs and restaurants, or for home use. The instruments, mostly called disc music boxes, were intentionally equipped with only a very few sound sources. Many of them only contained metal combs. For this reason they are also referred to as polyphones.

Although most disc music boxes offered a much lower sound quality than the orchestrions or the swiss music boxes they competed with, they soon became very successful due to their low price and the wide choice of music. The centre of production was Leipzig, were Lochmann started the production of disc music boxes under the trade name "Symphonion" immediately after he was granted the patent. Due to the large success, two of his employees founded the "Polyphon Musikwerke", also in Leipzig, in 1890. In the following years several hundreds of thousands instruments were made and exported to the whole world. The meteoric rise of this branch of industry abruptly ended after WW I when the gramophones started to conquer the markets.

CONCERT ORIGINAL.

Lochmann 300 „Concert Original“

- Plattengesteuerter Musikautomat mit Gewichtsantrieb
- Gebaut von Lochmann in Leipzig
- Maße (HxBxT): 280 cm x 115 cm x 58 cm
- Plattendurchmesser 74 cm
- Instrumente: Zwei Doppelkämme aus Metall, 12 Röhrenglocken, Mandolineneffekt und Schlaginstrumente
- Weltweit einziges Exemplar in dieser Bauart
- Unterschrank mit zehn Ersatzplatten
- Das „Concert Original“ markiert den Höhepunkt der plattengesteuerten mechanischen Musikinstrumente. Aufgrund des hohen Preises von 900 Reichsmark wurden aber nur wenige Exemplare gebaut.

- Disc Music Box with weight drive
- Built by Lochmann in Leipzig
- Dimensions (HxWxD): 280 cm x 115 cm x 58 cm
- Disc diameter 74 cm
- Instruments: Two metal double combs, 12 tubular chimes, mandoline effect and percussion
- Only specimen of this design worldwide
- There are ten spare discs in the lower part of the cabinet
- The „Concert Original“ marked a high spot of disc music boxes. However, due to its high price of 900 Reichsmark only a few specimens were built.

Oben: Der Instrumententeil des „Concert Original“. Rechts: Im Unterschrank befinden sich zehn Ersatzplatten.

Top: The instruments of the "Concert Original". Right: The lower part of the cabinet houses ten spare discs.

Lochmann Original

- Plattengesteuerter Musikautomat mit Federantrieb und Handaufzug
- Plattendurchmesser 62 cm
- Gebaut um 1905 von Lochmann in Leipzig
- Maße (HxBxT): 245 cm x 67,5 cm x 40 cm
- Instrumente: Metallkamm und Klangröhren
- Spezialausführung für den Export nach Übersee

- Disc music box with spring-wound clockwork
- Disc diameter 62 cm
- Built around 1905 by Lochmann in Leipzig
- Dimensions (HxWxD): 245 cm x 67.5 cm x 40 cm
- Instruments: Metal comb and tubular chimes
- Special design for export overseas

Produktion von Musikautomaten in der Fabrik von Paul Lochmann in Leipzig

Schlosserei-Maschinensaal.

Großuhrmacherei.

Die Herstellung der Gehäuse.

Das Ausstanzen der Notenscheiben.

Am 22. Oktober 1892 erschien in der „Illustrierte Zeitung" ein ausführlicher Bericht über die Musikautomatenproduktion in den Lochmann'schen Musikwerken in Leipzig-Gohlis. Die hier gezeigten Abbildungen sind diesem Artikel entnommen. Sie zeigen (von links oben im Uhrzeigersinn): den Maschinenraum der Schlosserei, die Produktion der Federantriebe, die Gehäuseproduktion und das Stanzen der Lochscheiben.

On 22. October 1892, a comprehensive article describing the production of automatic musical instruments in the Lochmann factory in Leipzig-Gohlis was published in the "Illustrierte Zeitung" ("Illustrated Newspaper"). The pictures shown here have been taken from this article. They show (clockwise from top left): the machine room of the locksmith's shop, production of the clockworks, production of the cabinets, and the punching of the metal discs.

Symphonion

- Plattengesteuerter Musikautomat mit Federantrieb und Handaufzug
- Plattendurchmesser 70 cm
- Gebaut von den Symphonion Musikwerken in Leipzig
- Maße (HxBxT): 249 cm x 112 cm x 61 cm
- Instrumente: Metall-Doppelkamm und 12 Glocken

- Disc music box with spring-wound clockwork
- Disc diameter 70 cm
- Built by the Symphonion Musikwerke in Leipzig
- Dimensions (HxWxD): 249 cm x 112 cm x 61 cm
- Instruments: Metal double comb and 12 bells

Symphonion

- Plattengesteuerter Musikautomat mit Federantrieb und Handaufzug und einem Doppelkamm aus Metall
- Plattendurchmesser 64 cm
- Gebaut von den Symphonion Musikwerken in Leipzig
- Maße (HxBxT): 248 cm x 85 cm x 51 cm

- Disc music box with spring-wound clockwork and a metal double comb
- Disc diameter 64 cm
- Built by the Symphonion Musikwerke in Leipzig
- Dimensions (HxWxD): 248 cm x 85 cm x 51 cm

Symphonion (links / left)

- Plattengesteuerter Musikautomat mit Federantrieb, Handaufzug und Metallkamm
- Plattendurchmesser 65 cm
- Gebaut von den Symphonion Musikwerken in Leipzig
- Maße (HxBxT): 245 cm x 81 cm x 47 cm

- Disc music box with spring-wound clockwork and metal comb
- Disc diameter 65 cm
- Built by the Symphonion Musikwerke in Leipzig
- Dimensions (HxWxD): 245 cm x 81 cm x 47 cm

Kalliope (rechts / right)

- Metallplattenspieler mit Federantrieb
- Plattendurchmesser 33,7 cm
- Gebaut von den Kalliope Musikwerken in Leipzig
- Maße (HxBxT): 19,5 cm x 41 cm x 41 cm
- Instrumente: Metallkamm und 10 Glocken

- Disc music box with spring-wound clockwork
- Disc diameter 33.7 cm
- Built by the Kalliope Musikwerke in Leipzig
- Dimensions (HxWxD): 19.5 cm x 41 cm x 41 cm
- Instruments: Metal comb and 10 bells

Polyphon Metallplattenspieler
Polyphon Disc Music Boxes

- Links: Metallplattenspieler mit Federantrieb und Handaufzug. Plattendurchmesser 40 cm. Maße (HxBxT): 34,5 cm x 54 cm x 49 cm
- Unten: Metallplattenspieler mit Federantrieb und Handaufzug. Plattendurchmesser 40 cm. Maße (HxBxT): 26 cm x 54 cm x 49 cm
- Beide mit einfachem Metallkamm

- Left: Disc music box with spring-wound clockwork. Disc diameter 40 cm. Dimensions (HxWxD): 34.5 cm x 54 cm x 49 cm
- Bottom: Disc music box with spring-wound clockwork. Disc diameter 40 cm. Dimensions (HxWxD): 26 cm x 54 cm x 49 cm
- Both with a single metal comb

Metallplattenspieler
Disc Music Boxes

- Links: Frühes Polyphon mit Vorderaufzug. Plattendurchmesser 28 cm. Maße (HxBxT): 20 cm x 34 cm x 30 cm.
- Unten: Schweizer Spieldose um 1890. Plattendurchmesser 11 cm. Maße (HxBxT): 13,5 cm x 18 cm x 18 cm.
- Beide mit Federantrieb und Handaufzug

- Left: Early Polyphon. The winding mechanism is located at the front. Disc diameter 28 cm. Dimensions (HxWxD): 20 cm x 34 cm x 30 cm.
- Bottom: Swiss disc music box from around 1890. Disc diameter 11 cm. Dimensions (HxWxD): 13.5 cm x 18 cm x 18 cm.
- Both with spring-wound clockwork

Gegenüberliegende Seite: Die Abbildung mit der lebensgroßen Mannequin-Puppe gibt einen Eindruck von der realen Größe der hier gezeigten Metallplattenspieler.

Opposite page: The picture with the life-size mannequin doll gives an impression of the real size of the disc music boxes shown here.

Polyphon Metallplattenspieler
Polyphon Disc Music Box

- Plattengesteuerter Musikautomat mit Federantrieb und Handaufzug
- Plattendurchmesser 56 cm
- Gebaut von den Polyphon Musikwerken in Leipzig
- Maße (HxBxT): 230 cm x 65 cm x 40 cm
- Instrumente: Metall-Doppelkamm und Stabglockenspiel

- Disc music box with spring-wound clockwork
- Disc diameter 56 cm
- Built by the Polyphon Musikwerke in Leipzig
- Dimensions (HxWxD): 230 cm x 65 cm x 40 cm
- Instruments: Metal double comb and bar chimes

Polyphon „Mikado“

- Plattengesteuerter Musikautomat mit Federantrieb und Handaufzug; Plattendurchmesser 62,5 cm; Doppelkamm aus Metall mit 159 Zähnen; Spieldauer bis zu 60 Minuten
- Gebaut von den Polyphon Musikwerken in Leipzig
- Maße (HxBxT): 218 cm x 86 cm x 53 cm
- Besonders reizvoll sind die Beleuchtung und die der berühmten Venus von Botticelli nachempfundene Figur, die sich zur Musik dreht

- Disc music box with spring-wound clockwork; disc diameter 62.5 cm; metal double comb with 159 teeth; playing time up to 60 minutes
- Built by the Polyphon Musikwerke in Leipzig
- Dimensions (HxWxD): 218 cm x 86 cm x 53 cm
- The illumination and the figure, modelled after the famous Venus by Botticelli, which is turning to the music, are particularly appealing

Standuhren mit Metallplattenspielern

Grandfather Clocks with Disc Music Boxes

- Links: Lenzkirch Uhrwerk von 1895 mit Monopol Plattenspieler. Plattendurchmesser 43 cm. Maße (HxBxT): 230 cm x 65 cm x 40 cm.
- Rechts: Lenzkirch Uhrwerk von 1901 mit Symphonion Plattenspieler. Plattendurchmesser 30 cm. Maße (HxBxT): 170 cm x 60 cm x 40 cm. Vier Alben mit Platten vorhanden.

- Left: Lenzkirch clockwork built 1895 with Monopol disc music box. Disc diameter 43 cm. Dimensions (HxWxD): 230 cm x 65 cm x 40 cm.
- Right: Lenzkirch clockwork built 1901 with Symphonion disc music box. Disc diameter 30 cm. Dimensions (HxWxD): 170 cm x 45 cm x 40 cm. Four albums with discs available.

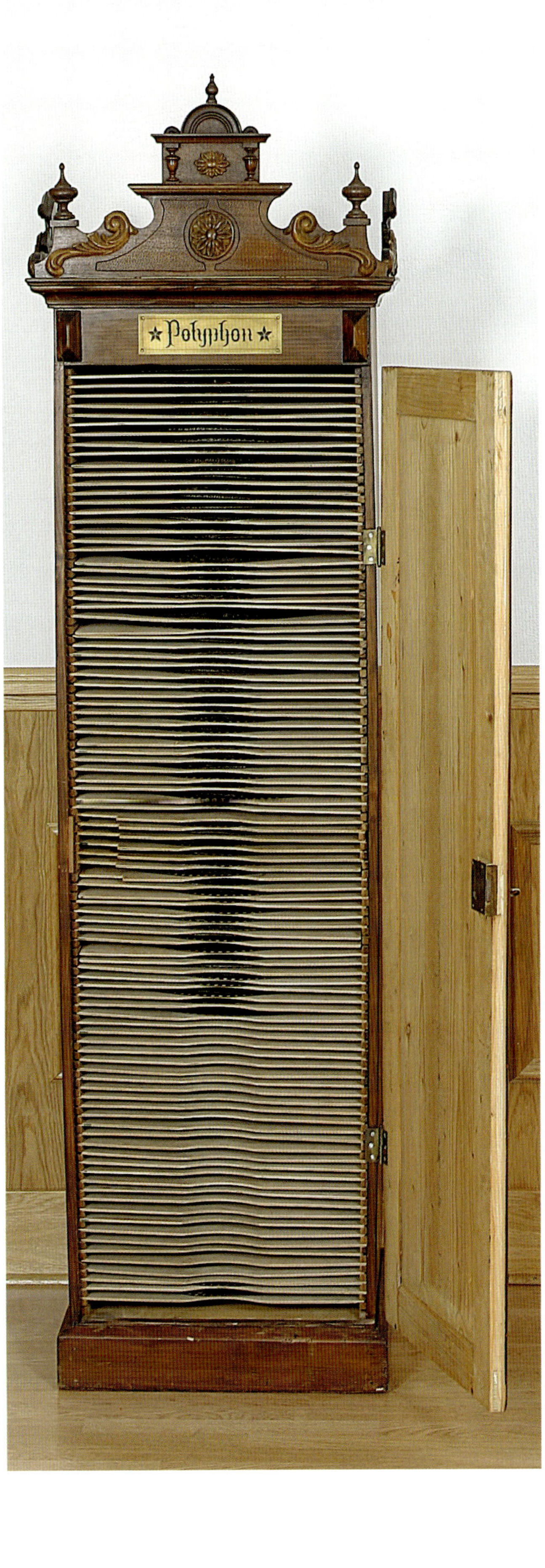

Polyphon Plattenschrank
Polyphon Disc Cabinet

- Schrank mit 120 Fächern zur Lagerung von Polyphon-Platten mit einem Durchmesser von 30 - 50 cm
- Maße (HxBxT): 192 cm x 50,5 cm x 48 cm

- Cabinet with 120 shelves for the storage of Polyphon discs with a diameter of 30 - 50 cm
- Dimensions (HxWxD): 192 cm x 50.5 cm x 48 cm

Metallplattenspieler - Disc Music Boxes

- Links: Symphonion „Lyra“ Musikautomat mit 30 cm Plattendurchmesser und Doppelkamm. Ein typisches Unterhaltungsgerät mit Münzeinwurf, das oft in Wartesälen zu finden war. Maße (HxBxT): 91 cm x 46 cm x 30 cm
- Rechts: Symphonion mit bewegten Figuren. Plattendurchmesser 33,5 cm. Maße (HxBxT): 69 cm x 41 cm x 30 cm
- Beide mit Federantrieb und Handaufzug. Die Figuren in der Mitte werden im Spieluhrenkapitel beschrieben.

- Left: Symphonion „Lyra“ disc music box with a disc diameter of 30 cm and double comb. A typical coin-operated instrument which was often placed in waiting rooms. Dimensions (HxWxD): 91 cm x 46 cm x 30 cm
- Right: Symphonion with moving figurines. Disc diameter 33.5 cm. Dimensions (HxWxD): 69 cm x 41 cm x 30 cm
- Both with spring-wound clockwork. The figures in the center are described in the music box chapter.

Chordephon

- Automatisch spielende Zither mit Drehkurbel. Durch einen komplizierten Anschlags- und Zupfmechanismus mit Dämpfung erzeugt das Gerät einen zitherähnlichen Klang. Typisch ist die außengezinkte Platte.
- Maße Instrument links (HxBxT): 18,5 cm x 96 cm x 78 cm. Plattendurchmesser 50,5 cm.
- Das Bild links zeigt ein in einen Tisch eingebautes Chordephon, das nach dem Einwurf eines Geldstücks gespielt werden konnte. Das Bild unten zeigt ein Gerät für den Heimgebrauch.

- Automatically playing zither with hand crank. It generates a zither like sound by a complicated pluck and strike mechanism with dampening. A typical feature is the plate with the teeth on the outer rim.
- Dimensions instrument left (HxWxD): 18.5 cm x 96 cm x 78 cm. Disc diameter 50.5 cm.
- The picture on the left shows a coin-operated chordephon built into a table. The picture at the bottom shows an instrument for use in the home.

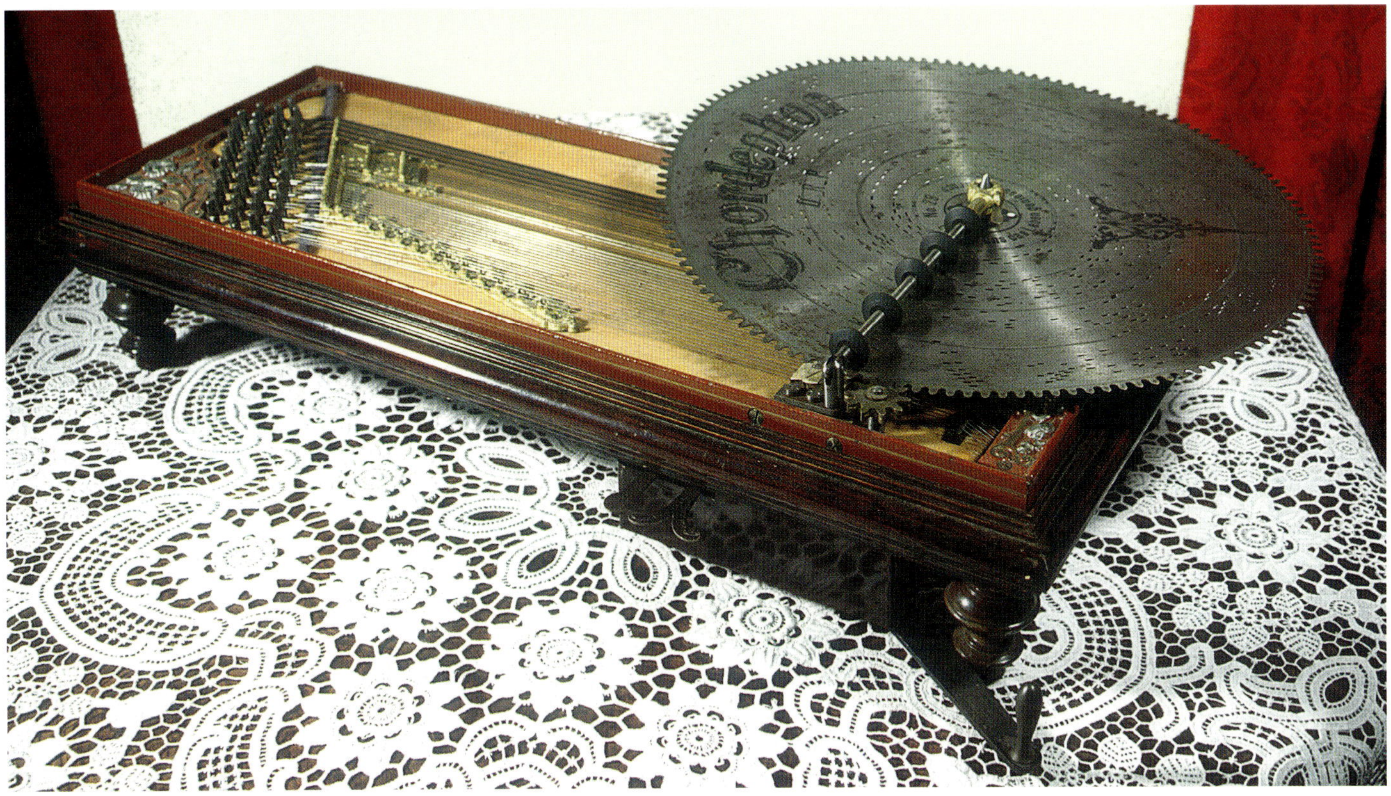

Die Sammlung Winkler
The Winkler Collection

Die Regierungszeit des letzten deutschen Kaisers Wilhelm II. von 1888 - 1918 ist als die Wilhelminische Ära in die deutsche Geschichte eingegangen. Es war genau diese, vom preußischen Militarismus geprägte Epoche, in der sich die plattengesteuerten Musikautomaten der größten Beliebtheit erfreuten. Einen faszinierenden Einblick in diese Zeit gibt die Sammlung Winkler im Wilhelmsbau. Sie umfasst eine einzigartige Kollektion von historischen Uniformen, Rangabzeichen, Orden, Helmen, Waffen und vielen weiteren Exponaten, die schwerpunktmäßig im Umfeld der ehemaligen württembergischen und preußischen Armee angesiedelt sind.

Begründet wurde die Sammlung vom Fabrikantenehepaar Erwin und Frida Winkler aus Cleebronn in Baden-Württemberg. In über 30-jähriger leidenschaftlicher Sammlertätigkeit wurde sie stetig ergänzt. So entstand schließlich im Laufe der Jahrzehnte eine einmalige und historisch wertvolle Dokumentation, die in Fachkreisen höchste Anerkennung gefunden hat.

Herr Winkler ist im Jahr 1993 an den Folgen einer schweren Krankheit verstorben. Wir sind glücklich und dankbar, dass die Sammlung nicht aufgelöst, sondern in ihrer Gesamtheit dem Museum schenkungsweise überlassen wurde, wo sie jetzt in ihrer ganzen Schönheit einer breiten Öffentlichkeit zugänglich ist und auf Dauer an die Stifter erinnern wird.

The reigning period of the last German emperor Wilhelm II. from 1888 - 1918 has made its way into history as the Wilhelminian Period. It was during this very epoch, largely influenced by Prussian militarism, when disc music boxes were most popular. The Winkler collection in the Wilhelmsbau offers a fascinating insight into this era of German history. It contains a unique collection of historic uniforms, badges and insignia, medals, helmets, weapons and many more exhibits with main emphasis on objects associated with the former armies of Württemberg and Prussia.

The collection was founded by the industrialist Erwin Winkler and his wife Frida of Cleebronn in Baden-Württemberg. In more than 30 years of passionate collecting it was constantly added to. In the course of decades these efforts resulted in a unique and historically valuable documentation highly acclaimed among experts.

Mr. Winkler died in 1993 from a severe illness. We are happy and grateful that the collection was not dissolved but donated in its entirety to the museum where it can now be shown to a wide public in all its beauty as a permanent memento of the donors.

Helme aus der links gezeigten Vitrine. Oben links: Helm eines Garde-Dragoners, Preußen 1871-1918. Oben rechts: Mütze eines Garde-Grenadiers, 1 Garde-Grenadier-Regiment, Preußen 1894. Unten rechts: Infanteriehelm, Preußen 1915.

Helmets from the left showcase. Top left: Helmet of a guard dragoon, Prussia 1871-1918. Top right: Cap of a guard grenadier, 1st Guard Granadier Regiment, Prussia 1894. Bottom right: Helmet of the infantry, Prussia 1915.

Gegenüberliegende Seite: General der württembergischen Armee. Bild links: Uniform eines württembergischen Hofbeamten.

Opposite page: General of the Württemberg Army. Picture left: Uniform of a courtier of Württemberg.

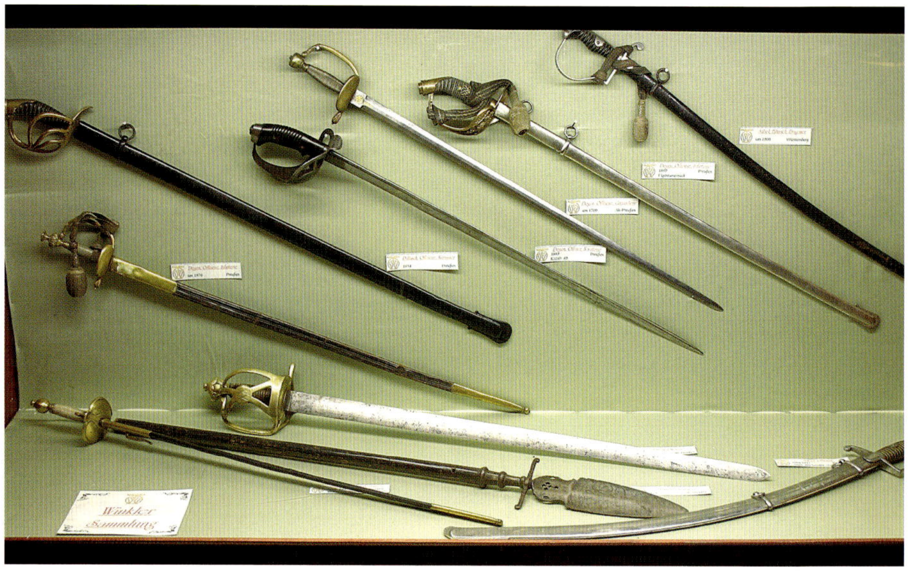
Winkler
Sammlung

Die Vitrinen mit Uniformen und historischen Waffen auf diesen beiden Seiten zeigen nur eine kleine Auswahl der in der Sammlung Winkler enthaltenen Exponate.

The showcases with uniforms and historical weapons on these two pages show only a small selection of the exhibits contained in the Winkler collection.

Prinz

Kirmes-, Kino- und Tanzorgeln

Von wenigen Ausnahmen wie der Welte Philharmonie Orgel oder der Aeolin Grand Konzertorgel abgesehen, wurden die großen, mechanisch gesteuerten Orgeln als Kirmesorgeln, Tanzorgeln oder Kinoorgeln eingesetzt. Die längste Tradition haben die Kirmesorgeln, die in der zweiten Hälfte des 19. Jahrhunderts aufkamen. Wie bei den Orchestrien nahm Deutschland auch beim Kirmesorgelbau eine führende Position ein. Das Zentrum der Produktion war Waldkirch im Schwarzwald, wo die beiden größten Hersteller A. Ruth & Sohn und die Fa. Bruder ihre Werkstätten hatten. Die frühen Instrumente wurden durch Stiftwalzen gesteuert, ab den 1890er Jahren verwendete man meist perforierte Kartonbänder (Kartonnoten), die wie ein Leporello stapelförmig gefaltet und auch als Notenbücher bezeichnet wurden. Die Instrumentierung der Kirmesorgeln bestand überwiegend aus Pfeifen und diversen Schlaginstrumenten. Die große Kunst des Kirmesorgelbaus bestand darin, preiswerte Instrumente zu bauen, die gut klangen und eine hohe Lautstärke besaßen.

Aus den Kirmesorgeln entwickelten sich die Tanzorgeln, die größten mechanischen Musikinstrumente, die jemals gebaut wurden. Auch hier werden gefaltete Kartonbänder als Tonträger verwendet. Die beiden bedeutendsten Hersteller waren Mortier und Decap, beide in Antwerpen / Belgien. Während der Orchestrionbau Ende der 1920er Jahre ein abruptes Ende fand, wurden Tanzorgeln noch nach dem 2. Weltkrieg in bedeutenden Stückzahlen gebaut. Bei Mortier endete die Produktion erst 1952. Die Firma Decap, die die Reste von Mortier übernahm, existiert noch heute und baut weiterhin Tanzorgeln.

Zur musikalischen Untermalung von Stummfilmen begnügte man sich anfangs mit einem Klavier. Um die dramatischen Effekte akustisch besser darstellen zu können, begannen einzelne Hersteller spezielle Kinoorgeln zu bauen. Diese verfügten über spezielle Toneffekte, mit denen die unterschiedlichsten Geräusche wie herannahende Autos, Donner, Regen, Vogelgezwitscher usw. imitiert werden konnten. Auch die Firma Welte hat solche Kinoorgeln gebaut. Ein bedeutender Hersteller in den USA war die Fa. Wurlitzer, die später mit Jukeboxen Furore machen sollte. Mit dem Aufkommen des Tonfilms 1929 endete der Bedarf an solchen Instrumenten.

Fairground-, Cinema-, and Dance Organs

With few exceptions like the Welte Philharmonie Organ or the Aeolian Grand concert organ, the large automatic organs were used as fairground organs, dance organs, or cinema organs. Fairground organs, which originated in the second half of the 19th century, have the longest tradition. Germany had a leading position not only in building orchestrions but also in building fairground organs. The center of production was Waldkirch in the Black Forest where the two largest producers, A. Ruth & Son and the Bruder company, were based. The early instruments were controlled by pinned barrels. Starting around 1890, perforated cardboard strips became dominant. They were folded zig-zag fashion into a compact pile referred to as "book". Therefore the term "book music". As sound sources the fairground organs mostly contained pipes and percussion. The art of fairground organ manufacturing was to build powerful well-sounding instruments at an affordable price.

The fairground organs provided the basis for the development of the dance organs, the largest automatic musical instruments ever built. The two most important manufacturers were Mortier and Decap both based in Antwerp / Belgium. Most dance organs also play book music. Whereas orchestrion manufacturing abruptly ended at the end of the 1920s, dance organs were built in significant numbers long after WW II. Mortier produced dance organs until 1952. Decap, which took over the remains of Mortier, still exists and continues to build dance organs.

At the beginning a piano was sufficient for the accompaniment of silent movies. Then some manufacturers started to produce special cinema organs, also called photoplayers, that delivered a better musical background for the dramatic effects. They contained special sound effects for the imitation of many different sounds like approaching cars, thunder, rain, bird chirping etc. The Welte company in Freiburg also built cinema organs. A major manufacturer in the US was the Wurlitzer company which later became famous for their jukeboxes. After the advent of the sound movie in 1929, the demand for cinema organs ended.

Rosskopf's
Etagenkarussell
Schifferstadt
Pfalz

Etagenkarussell mit Bruder Orgel
Two Floor Carousel with Bruder Organ

- Hervorragend erhaltenes, historisches Etagenkarussell gebaut um 1890 mit einer Höhe von 7 m und einem Durchmesser von 11 m
- 20 geschnitzte Holzpferde in der unteren Etage und 4 Schiffe aus Metall mit Wippen in der oberen Etage
- Aufwändige Holzfassade mit einer fantastischen Beleuchtung
- Für die richtige Musik sorgt eine ca. 150 Jahre alte Orgel der Fa. Bruder / Waldkirch im Schwarzwald mit Orgelpfeifen, Bassposaunen, Xylophon und Schlagzeug.
- Die Orgel war ursprünglich mit einer Stiftwalze ausgestattet und wurde später mit einer 33er Ruth Kartonbandsteuerung nachgerüstet
- Das Karussell, das viele Jahre von der Familie Otto Rosskopf aus Schifferstadt betrieben wurde, ist nicht nur ein musikalisches und nostalgisches Juwel sondern auch ein technisches Glanzstück vergangener Zeiten

- Historical two floor carousel built around 1890 in excellent condition with a height of 7 meters and a diameter of 11 meters
- 20 carved wooden horses in the lower floor and 4 teeter totters with metal ships in the upper floor
- Sophisticated facade with a magnificent illumination
- The right music is provided by a 150 years old organ made by Bruder / Waldkirch in the Black Forest equipped with pipes, bass trombones, xylophone, and percussion
- The organ was originally equipped with a pinned barrel and was later remodeled with a 33 keys Ruth key frame
- The carousel, run for many years by Otto Rosskopf of Schifferstadt and his family, is not only a musical and nostalgic jewel but also a technical masterpiece of bygone times

Bruder Kirmesorgel
Bruder Fairground Organ

- Gebaut 1905 von Bruder in Waldkirch
- Maße (HxBxT): 235 cm x 240 cm x 90 cm
- Spielt Papierrollen Modell 111, 43 Claves, Breite 200 mm
- Intrumente: Pfeifen und Schlagzeug

- Built 1905 by Bruder in Waldkirch
- Dimensions (HxWxD): 235 cm x 240 cm x 90 cm
- Plays paper rolls model 111, 43 keys, width 200 mm
- Instruments: Pipes and percussion

Molzer Kirmesorgel
Molzer Fairground Organ

- Gebaut 1920 von Molzer in Wien
- Maße (HxBxT): 200 cm x 310 cm x 95 cm
- Spielt Kartonnoten, 46er Skala
- Intrumente: Pfeifen und Schlagzeug
- Molzer war ein bekannter Orgelbauer, der u.a. für den Wiener Prater ein Großorchestrion fertigte

- Built 1920 by Molzer in Vienna
- Dimensions (HxWxD): 200 cm x 310 cm x 95 cm
- Plays music books with 46 keys
- Instruments: Pipes and percussion
- Molzer was a renowned organ manufacturer who also built a large orchestrion for the Vienna Prater

Ruth Walzenorgel (links)
Ruth Barrel Organ (left)

- Gebaut 1890 von Ruth in Waldkirch
- Maße (HxBxT): 170 cm x 120 cm x 70 cm
- Walzenbreite 113 cm
- 49 Claves, 4 Register

- Built 1890 by Ruth in Waldkirch
- Dimensions (HxWxD): 170 cm x 120 cm x 70 cm
- Barrel width 113 cm
- 49 keys, 4 registers

Loos Orgel (rechts)
Loos Organ (right)

- Gebaut 1900 von Loos in Nürnberg
- Maße (HxBxT): 172 cm x 120 cm x 68 cm
- Spielt Kartonnoten Breite 175 mm

- Built 1900 by Loos in Nuremberg
- Dimensions (HxWxD): 172 cm x 120 cm x 68 cm
- Plays music books with a width of 175 mm

Mortier Kaffeehaus Orgel - Mortier Coffee House Organ

- Gebaut 1925 von Mortier in Antwerpen / Belgien
- Maße (HxBxT): 288 cm x 314 cm x 175 cm
- Spielt Kartonnoten mit einer Breite von 305 mm
- Orgel mit 86 Tonstufen
- Besonders hochwertige Tanzorgel aus edlen Hölzern gefertigt mit aufwändig verzierter Fassade und sich drehender Figur
- Über 1200 Notenbücher mit Unterhaltungsmusik vorhanden

- Built 1925 by Mortier in Antwerp / Belgium
- Dimensions (HxWxD): 288 cm x 314 cm x 175 cm
- Plays music books with a width of 305 mm
- Organ with 86 keys
- A particularly high-class dance organ made of high-quality woods with an extravagantly decorated front and a turning figure
- More than 1,200 music books with light music available

Morton Kinoorgel - Morton Cinema Organ

- Gebaut 1925 von der Firma Morton / USA
- Maße (HxBxT): 307 cm x 320 cm x 186 cm
- Spielt Papierrollen mit einer Breite von 287 mm
- Die zweimanualige Orgel mit 88 Tonstufen kann sowohl von Hand als auch automatisch gespielt werden. Spektakulär sind die Beleuchtungseffekte, die das Spiel der Orgel begleiten.
- Die Orgel hat zwei Rollen, die wechselseitig genutzt werden und so ein unterbrechungsfreies Spiel ermöglichen. Sie hat 12 Tasten für Toneffekte wie Kanonendonner, Vogelgezwitscher, Gewitter etc.

- Built 1925 by Morton / USA
- Dimensions (HxWxD): 307 cm x 320 cm x 186 cm
- Plays paper rolls with a width of 287 mm
- The Organ with two manuals and 88 keys can be played either automatically or by hand. A spectacular feature are the illumination effects which accompany the play of the organ.
- The organ is equipped with two rolls that are used alternately and allow a continuous play without interruption. It has 12 keys for sound effects like canon thunder, bird chirping, thunderstorm etc.

Mortier Tanzorgel - Mortier Dance Organ

- Gebaut 1947 von Mortier in Antwerpen / Belgien
- Maße (HxBxT): 2,80 Meter x 4,10 Meter x 2,70 Meter
- Die Orgel spielt Kartonnoten mit einer Breite von 448 mm
- Der Orgelteil der Tanzorgel umfasst 125 Tonstufen. Er wird von drei Akkordeons und einem Schlagzeug unterstützt. Als Blickfang dient eine bunte elektrische Beleuchtung.
- Das Instrument wurde ursprünglich für einen amerikanischen Auftraggeber gebaut. Es kam 1993 in das Museum Speyer und wurde im Stil der damaligen Zeit renoviert.

- Built 1947 by Mortier in Antwerp / Belgium
- Dimensions (HxWxD): 2.80 meters x 4.10 meters x 2.70 meters
- Plays music books with a width of 448 mm
- The organ part of the Instrument has 125 keys. It is supported by three accordions and percussion. A colorful illumination is used as eyecatcher.
- The instrument was originally been built for an American customer. In 1993 it came into the museum and was renovated in the style of its time.

Decap Tanzorgel - Decap Dance Organ

- Gebaut in den 1940er Jahren von Decap in Antwerpen / Belgien
- Maße (HxBxT): 2,70 Meter x 4,35 Meter x 2,00 Meter
- Die Orgel spielt Kartonnoten mit einer Breite von 330 mm
- Neben dem Orgelteil verfügt das Instrument über zwei spielende Akkordeons, ein imitiertes Saxophon und ein Schlagzeug. Während des Spiels dreht sich das Instrument.

- Built in the 1940s by Decap in Antwerp / Belgium
- Dimensions (HxWxD): 2.70 meters x 4.35 meters x 2.00 meters
- Plays music books with a width of 330 mm
- Besides the organ part the instrument is equipped with two playing accordions, an imitated saxophone, and percussion. The organ turns while it is playing.

Decap Tanzorgel - Decap Dance Organ

- Gebaut um 1960 von Decap in Antwerpen / Belgien
- Maße (HxBxT): 2,60 Meter x 4,70 Meter x 1,70 Meter
- Die Orgel spielt Kartonnoten mit einer Breite von 375 mm
- Instrumente: Zwei spielende Akkordeons mit 41 Pianotasten und 120 Basstasten, ein Saxophon, eine Horntrompete, diverse Pfeifen, Schlagzeug und elektronische Klangerzeuger

- Built around 1960 by Decap in Antwerp / Belgium
- Dimensions (HxWxD): 2.60 Meter x 4.70 Meter x 1.70 Meter
- Plays music books with a width of 375 mm
- Instruments: Two playing accordions with 41 piano keys and 120 bass keys, one saxophone, one horn trumpet, various pipes, percussion, and electronic sound producers

Decap Tanzorgel - Decap Dance Organ

- Gebaut 1969 von Decap in Antwerpen / Belgien
- Maße (HxBxT): 3,80 Meter x 6,60 Meter x 2,20 Meter
- Die Orgel spielt Kartonnoten mit einer Breite von 375 mm
- Instrumente: Zwei spielende Akkordeons mit 41 Pianotasten und 120 Basstasten, zwei Reihen Holzpfeifen, ein Saxophon, Schlagzeug und elektronische Klangerzeuger

- Built 1969 by Decap in Antwerp / Belgium
- Dimensions (HxWxD): 3.80 meters x 6.60 meters x 2.20 meters
- Plays music books with a width of 375 mm
- Instruments: Two playing accordions with 41 piano keys and 120 bass keys, two sets of wooden pipes, one saxophone, percussion, and electronic sound producers

Mortier Tanzorgel

- Gebaut um 1922 von Mortier in Antwerpen / Belgien
- Mit einer Fassadenbreite von 12,5 Metern ist diese Orgel die breiteste Tanzorgel der Welt. Das Gewicht beträgt 2,8 Tonnen, die Höhe 2,90 Meter und die Tiefe 1,10 Meter.
- Die Orgel spielt Kartonnoten mit einer Breite von 360 mm
- Die Orgel kam vermutlich zunächst in Belgien zum Einsatz. Später gelangte sie nach Paris, wo sie unfachmännisch umgebaut wurde. Die Orgel wurde vom Museum in einem äußerst schlechten Zustand erworben und von 2001 bis 2004 vom Orgelbaumeister Johann Gebert mit einem Zeitaufwand von rund 6000 Stunden komplett erneuert.
- Die künstlerische Ausarbeitung der Bildtafeln und der Figuren erfolgte durch die Mutter des Orgelbauers, der renommierten Künstlerin und Märchenbuchillustratorin Helga Gebert
- Große Teile der Fassade wurden mit Blattgold überzogen, die 11 Figuren sind mit einer Bewegungsmechanik versehen
- Die Orgel hat 101 Tonstufen, 16 Register, 333 Pfeifen mit bis zu drei Metern Länge, 2 große Trommeln, 2 kleine Trommeln, einen Holzblock, Triangel und Becken
- Aufgrund der großen Skala ist die Orgel fähig, jede Musik wiederzugeben, von großen Ouvertüren bis hin zum Rock'n Roll

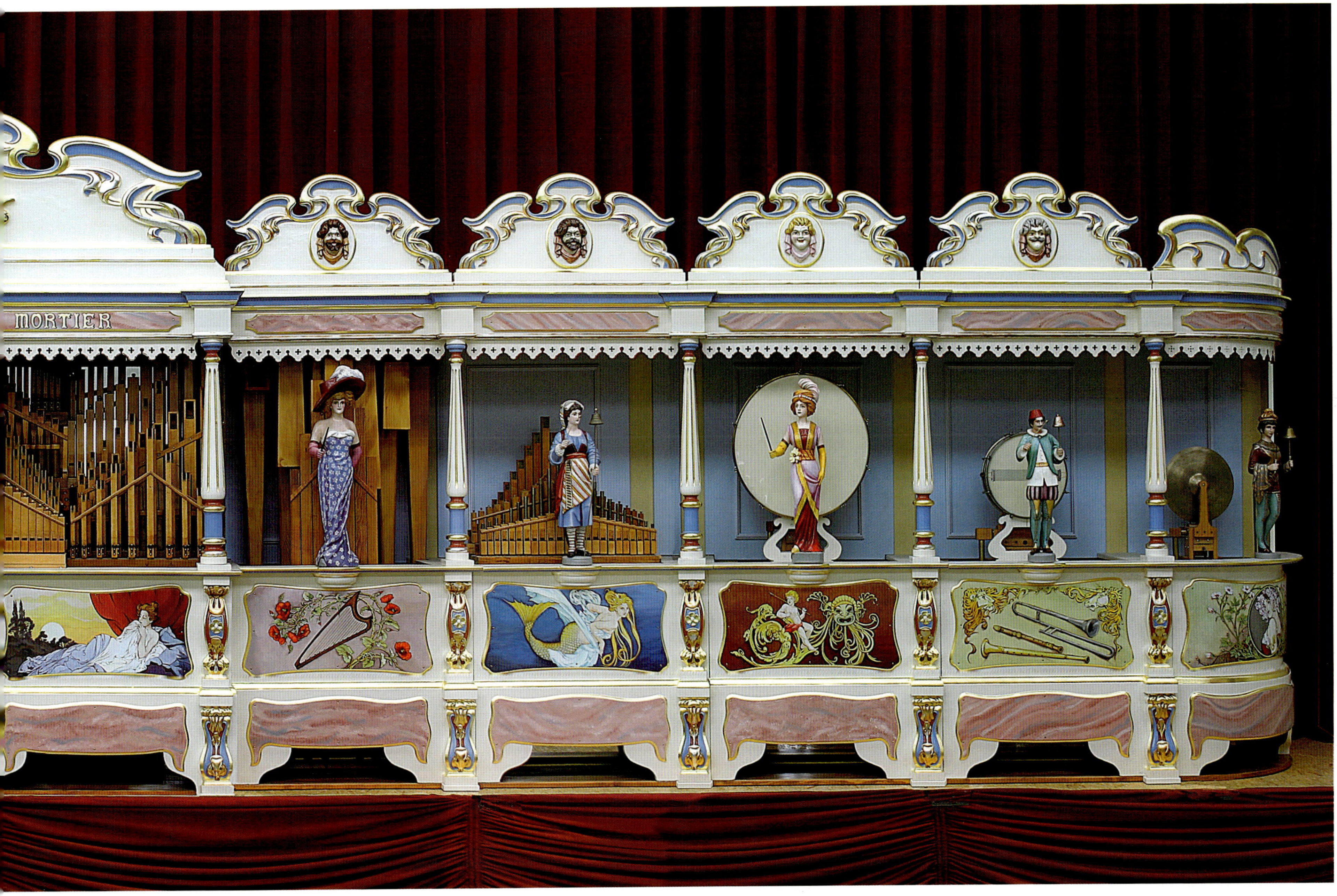

Mortier Dance Organ

- Built around 1922 by Mortier in Antwerp / Belgium
- With a front width of 12.5 meters it is the widest dance organ in the world. It weighs 2.8 tons and has a height of 2.90 meters and a depth of 1.10 meters.
- The organ plays music books with a width of 360 mm
- The organ was probably first used in Belgium. It later came to Paris were it was remodeled in a non-professional way. The organ was acquired by the museum in a very poor condition. Between 2001 and 2004 it was completely renovated by the organ builder Johann Gebert with an effort of more than 6,000 working hours.
- The painting of the front and the figures was done by the mother of the organ builder, the renowned artist and illustrator of fairy tale books Helga Gebert
- Large areas of the front are covered with gold leaf, the 11 figures are animated using a special mechanics
- The organ has 101 keys, 16 registers, 333 pipes with a length of up to 3 meters, two large drums, 2 small drums, one wood block, triangle, and cymbals
- Due to the large musical scale the organ can play all types of music ranging from classic ouvertures to rock'n roll

THEOPHILE MORTIER

Gegenüberliegende Seite: Die Mortier Tanzorgel in der Liller Halle des Technik Museum Speyer. Bilder oben: Die künstlerische Gestaltung der Orgel erfolgte durch die renommierte Künstlerin und Märchenbuchillustratorin Helga Gebert. Darunter: Große Teile der Fassade wurden mit 22 Karat Blattgold vergoldet. Insgesamt wurden 1800 Blatt Gold und Silber aufgetragen, was einer Fläche von 11 Quadratmetern aber nur 30 Gramm Material entspricht. Bild links: Orgelbaumeister Johann Gebert mit einer der großen Orgelpfeifen.

Opposite page: The Mortier dance organ in the Liller Hall of the Technik Museum Speyer. Pictures on top: The paintings were done by Helga Gebert, renowned artist and illustrator of fairy tale books. Below: Large areas of the front were covered with 22-carat gold leaves. In total, 1800 leaves of gold and silver were put on which corresponds to an area of 11 sqm but a mere 30 grams of material. Picture left: Organ builder Johann Gebert with one of the large organ pipes.

Drehorgeln, Organios und Straßenklaviere

Neben den großen Standgeräten entstanden schon früh mobile Musikautomaten. Die ältesten erhaltenen Geräte stammen aus der Zeit um 1800. Beim Instrumententeil können im Wesentlichen zwei Bauformen unterschieden werden. Bei den sogenannten Straßenklavieren besteht die Instrumentierung aus einem Klavierteil mit Schlaginstrumenten. Eng verwandt mit diesen mobilen Instrumenten sind die Organios, die in gleicher Weise instrumentiert sind, aber oft als Standgeräte genutzt wurden. Sie waren in Frankreich und Italien sehr populär. Die in Deutschland weit verbreiteten Drehorgeln sind dagegen mit Metall- oder Holzpfeifen ausgestattet. Daneben produzierten die Musikautomatenhersteller noch zahlreiche weitere preiswerte Kurbelinstrumente. Einige Beispiele finden sich am Ende dieses Kapitels.

Die größten Drehorgelfabriken befanden sich in Berlin und in Waldkirch / Schwarzwald. Berühmte Firmen waren u.a. Bacigalupo, Frati und Bruder. Wie bei den Großinstrumenten erfolgte die Steuerung bei den trag- bzw. fahrbaren Geräten anfangs über eine Stiftwalze. Später wurden auch Kartonbänder oder Papierrollen verwendet. Als Antrieb diente eine Handkurbel. Manche Straßenklaviere verfügen auch über einen Federantrieb.

Bis auf wenige Ausnahmen dienten die Kurbelinstrumente dem fahrenden Volk sowie Straßenmusikanten und Bettlern zum Broterwerb. Da sich die wenigsten ein solches Instrument kaufen konnten, entwickelte sich in den Städten ein reger Drehorgelverleih. Der so schon recht karge Verdienst der Drehorgelspieler wurde durch die hohen Leihgebühren noch weiter gemindert und reichte kaum zum Überleben.

Hand-cranked Organs, Organios and Street Pianos

Besides the large stationary devices there have always been mobile automatic musical instruments. The earliest specimens that survived were built around 1800. Based on instrumentation, two designs can be distinguished. The so-called street pianos are equipped with a piano part and percussion. They are closely related to the organios which contain the same instruments but were often intended as stationary devices. Both instrument types were popular in Italy and France. On the other hand, hand-cranked organs, which were wide-spread in Germany, are equipped with different sorts of wooden and metal pipes. Besides these many other cheap hand-cranked instruments were built. Some examples are shown at the end of this chapter.

The largest manufacturers of hand-cranked organs were based in Berlin and Waldkirch / Black Forest. Famous companies were Bacigalupo, Frati and Bruder. They first used pinned barrels as sound carriers that were later substituted by perforated cardboard or paper bands. Most instruments were hand-cranked, some street pianos were equipped with spring-wound clockwork.

With few exceptions, hand-cranked instruments were used by itinerants, street musicians and beggars to make a living. Since most of them could not afford to buy one a brisk trade of organ rentals began to develop in the cities. This way, the earnings of a hurdy-gurdy man, which were quite meager to start with, were further diminished by high rental fees and therefore hardly enough to survive.

Spanisches Straßenklavier
Spanish Street Piano

- Gebaut um 1880 von Luis Casali in Barcelona
- Maße (HxBxT): 156 cm x 123 cm x 61 cm
- Das Straßenklavier ist das Gegenstück zur Drehorgel. Das hier gezeigte walzengesteuerte Instrument (Walzenbreite 102 cm) beherrscht zehn Musikstücke, die von Hand vorgewählt werden können.

- Built around 1880 by Luis Casali in Barcelona
- Dimensions (HxWxD): 156 cm x 123 cm x 61 cm
- The street piano is the counterpart of the hand-cranked organ. The instrument with pinned barrel (barrel width 102 cm) shown here plays ten tunes which can be manually pre-selected.

Italienisches Straßenklavier
Italian Street Piano

- Größtes spielbereites italienisches Straßenklavier
- Maße (HxBxT): 274 cm x 174 cm x 63 cm
- Als Tonträger dienen drei Stiftwalzen mit einer Breite von 154 cm. Das musikalische Repertoire dieses extrem seltenen Instruments ist einzigartig.
- Um das Publikum anzulocken, wurden die Straßenklaviere oft mit pompösen Aufbauten versehen. Das linke Bild zeigt das Instrument mit und das untere Bild ohne einen solchen Zusatz.

- Large functional Italian street piano
- Dimensions (HxWxD): 274 cm x 174 cm x 63 cm
- As sound carriers serve three pinned barrels with a width of 154 cm. The music repertoire of this very rare instrument is unique.
- To attract an audience, street pianos were often decorated with pompous tops. The picture left shows the instrument with and the picture below without the top.

Italienisches Straßenklavier (links) Italian Street Piano (left)

- Kleines italienisches Straßenklavier von 1870
- Auf der Walze mit einer Breite von 51 cm befinden sich sechs Musikstücke
- Maße (HxBxT): 137 cm x 68 cm x 43 cm

- Small Italian street piano built around 1870
- The barrel with a width of 51 cm plays six pieces of music
- Dimensions (HxWxD): 137 cm x 68 cm x 43 cm

Piano Melodico (rechts / right)

- Ein sehr populäres, handbetriebenes Instrument insbesondere für den Heimgebrauch gedacht, das zwischen 1890 und 1910 zehntausendfach verkauft wurde. Als Klangerzeuger dienen 54 Klaviersaiten die, gesteuert durch ein Kartonband mit einer Breite von 273 mm, über einen Hebelmechanismus angeschlagen werden. Der Klang ähnelt dem einer Zither. Maße (HxBxT): 107 cm x 84 cm x 42 cm

- A highly popular hand-cranked instrument, particularly for home use. Tens of thousands of this model were sold between 1890 and 1910. The sound sources used were 54 piano strings which, controlled by a perforated cardboard with a width of 273 mm, were touched via lever mechanisms. The sound was similar to that of a zither. Dimensions (HxWxD): 107 cm x 84 cm x 42 cm.

Organio

- Prunkvolles, französisches Organio von 1882 mit wertvollen Schnitzereien
- Instrumente: Klavier, Mandoline, Kastagnetten, Becken, Triangel und Trommeln
- Maße (HxBxT): 233 cm x 153 cm x 68 cm

- High-quality French organio from 1882 with elaborate carvings
- Instruments: Piano, mandoline, castanets, cymbal, triangle, and drums
- Dimensions (HxBxT): 233 cm x 153 cm x 68 cm

Organio

- Einfaches, französisches Organio mit Federantrieb von 1893
- Maße (HxBxT): 131 cm x 82 cm x 54 cm
- Walzenbreite 61 cm
- Instrumente: Klavier und Trommel

- Simple French organio with spring-wound clockwork built 1893
- Dimensions (HxWxD): 131 cm x 82 cm x 54 cm
- Barrel width 61 cm
- Instruments: Piano and drum

Organio

- Einfaches Organio mit Federantrieb
- Maße (HxBxT): 190 cm x 123 cm x 71 cm
- Instrumente: Klavier, kleine Trommel, Triangel, Kastagnetten

- Simple organio with spring-wound clockwork
- Dimensions (HxWxD): 190 cm x 123 cm x 71 cm
- Instruments: Piano, small drum, triangle, castanets

Drehorgeln
Hand-cranked Organs

- Oben: 18er Umhängeorgel von Bacigalupo Söhne. Maße (HxBxT): 36,5 cm x 34,5 cm x 25 cm.
- Links: 20er Walzendrehorgel mit Gedacktflöte gebaut von Kamenik in Prag. Walzenbreite 28 cm. Maße (HxBxT): 43,5 cm x 38 cm x 25 cm.

- Top: Shoulder organ with 18 keys by Bacigalupo Sons. Dimensions (HxWxD): 36,5 cm x 34,5 cm x 25 cm.
- Left: Hand-cranked barrel organ with "Gedacktflöte" built by Kamenik in Prague. Barrel width 28 cm. Dimensions (HxWxD): 43,5 cm x 38 cm x 25 cm.

Audio CD Titel 28 + 29

Drehorgeln - Hand-cranked Organs

- Oben links: 33er Drehorgel gebaut 1925 von Frati in Berlin. Maße (HxBxT): 113 cm x 57 cm x 35 cm.
- Oben rechts: Walzendrehorgel mit 33 Tonstufen, gebaut 1980 von Arnold in Bad Schönborn. Spielt 13 Musikstücke. Gewicht 62,5 kg. 213 Pfeifen (Trompete, Pan und Violine) in sieben Registern. Maße (HxBxT): 115 cm x 82 cm x 52 cm.
- Unten links: Walzengesteuerte 48er Trompetendrehorgel mit Piccolo gebaut um 1900 von Bacigalupo & Söhne in Berlin. Maße (HxBxT): 127 cm x 78 cm x 48 cm.

- Top left: Hand-cranked organ with 33 keys built in 1925 by Frati in Berlin. Dimensions (HxWxD): 113 cm x 57 cm x 35 cm.
- Top right: Hand-cranked barrel organ with 33 keys built in 1980 by Messrs. Arnold / Bad Schönborn. Plays 13 tunes. Weight 62.5 kg. 213 pipes (trumpet, pan and violin), seven stops. Dimensions (HxWxD): 115 cm x 82 cm x 52 cm.
- Bottom left: Hand-cranked 38 keys trumpet barrel organ with piccolo built around 1900 by Bacigalupo & Sons in Berlin. Dimensions (HxWxD): 127 cm x 78 cm x 48 cm.

Phönix Zungenorgel
Phönix Organette (Reed Organ)

- Gebaut um 1900 von den Schmidt & Co. Phönix Musikwerke in Leipzig. Mit der Handkurbel wird eine Luftpumpe betätigt und der Tonträger, ein perforierter Metallring aus Zink, gedreht. Ein billiges Instrument, das in großen Stückzahlen gebaut wurde und Bettlern oft zum Broterwerb diente. Maße (HxBxT): 44,5 cm x 34 cm x 23 cm.

- Built around 1900 by the Schmidt & Co. Phönix Musikwerke in Leipzig. The hand crank drives an air pump in the instrument and turns the sound carrier, a perforated metal ring made of zinc. A cheap instrument produced in large numbers which was often used by beggars to make a living. Dimensions (HxWxD): 44.5 cm x 34 cm x 23 cm.

Manopan Zungenorgel - Manopan Organette (Reed Organ)

- Gebaut um 1900 von den Euphonika Musikwerken in Leipzig. Durch die Kurbel wird das perforierte Kartonband bewegt und eine Luftpumpe betätigt. Wenn eine der Öffnungen im Kartonband den Lesekopf passiert, bringt der Luftstrom die Metallzungen zum Schwingen. Auf der Oberseite ist das Gerät mit einer kunstvollen Widmung versehen.
- Maße (HxBxT): 33,5 cm x 65,5 cm x 38 cm
- Breite des Kartonbands 165 mm

- Built around 1900 by the Euphonika Musikwerke in Leipzig. The hand crank moves the perforated cardboard strip and drives an air pump in the instrument. Whenever one of the perforations in the cardboard strip passes over the tracker bar the resulting airflow causes the reeds to vibrate. The top surface of the instrument is adorned with an intricate dedication.
- Dimensions (HxWxD): 33.5 cm x 65.5 cm x 38 cm
- Width of the cardboard strip 165 mm

Spielzeugorganio - Toy Organio

- Nachbildung eines Organios als Kinderspielzeug
- Maße (HxBxT): 45 cm x 40 cm x 28 cm

- Replica of an organio for use as a toy
- Dimensions (HxWxD): 45 cm x 40 cm x 28 cm

Triola

- Teilautomatische Zither. Die Melodie wird über die per Handkurbel betätigte Rolle gespielt. Die Begleitung muss manuell erfolgen. Die Akkorde sind aber auf der Rolle angegeben (jetzt Akkord 1, dann 4, dann 2 etc.). Maße (HxBxT): 14 cm x 42 cm x 55 cm. Breite der Papierrolle 150 mm.

- Partially automatic zither. The Melody is played by the hand-cranked music roll. The accompaniment must be played manually. The chords are noted on the roll (now chord 1, then 4, then 2 etc.). Dimensions (HxWxD): 14 cm x 42 cm x 55 cm. Width of the paper roll 150 mm.

Cabinet Roller Organ

- Gebaut um 1890 von der Autophone Co. Ithaka / USA
- Kleine Tischorgel mit spiralförmig bestifteter Walze und 20 Zungen. Auch als „Methodisten Orgel" bezeichnet, da sie vorwiegend für religiöse Lieder verwendet wurde.
- Maße (HxBxT): 46 cm x 38 cm x 32 cm

- Built around 1890 by the Autophone Co. Ithaka / USA
- Small table organ with spiral pinned barrel and 20 reeds. Also called "Methodist Organ" since it was mostly used for religious songs.
- Dimensions (HxWxD): 46 cm x 38 cm x 32 cm

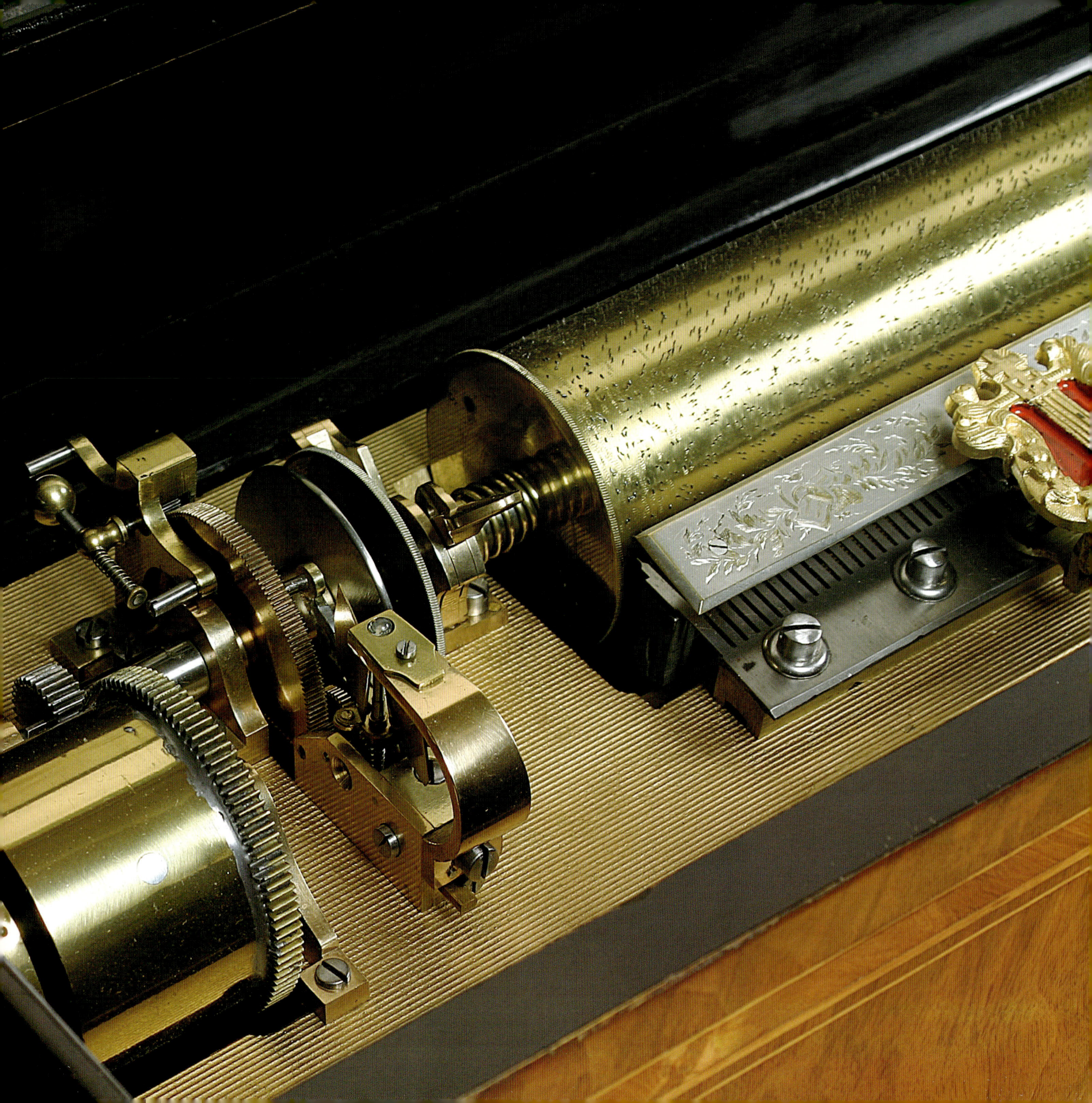

Musikdosen und Spieluhren

Neben Pfeifen gehörten Metallkämme zu den ersten Klangerzeugern, die in Musikautomaten Verwendung fanden. Erfunden wurde der Metallkamm als Tonquelle um 1796 vom Schweizer Uhrmacher Antoine Favre in Genf. Die ersten Spielwerke mit Metallkamm wurden in Taschenuhren eingebaut, wodurch die Bezeichnung „Spieluhr" entstand. In der Folgezeit entwickelte sich in der Schweiz eine sehr erfolgreiche Spieluhrenindustrie, die insbesondere im Genfer Raum angesiedelt war. Aus einfachen Spielwerken mit Kämmen, die weniger als zehn Zähne hatten, wurden Wunderwerke der Feinmechanik, die in kostbare Gehäuse eingebaut und als Spieldosen bezeichnet wurden. Neben dem Metallkamm enthalten viele Spieldosen weitere Klangerzeuger wie kleine Trommeln, Glocken oder durchschlagende Zungen. Der bedeutendste Hersteller war Nicole Frères in Genf, in Deutschland produzierte die Firma Lamy in Furtwangen hochwertige Spieldosen.

Alle klassischen Spieldosen verwenden Stiftwalzen als Tonträger. Die größten tragen mehrere zehntausend Stifte, was aber dennoch nur für wenige Musikstücke reicht. Mit der Erfindung der preiswerten Lochscheibe als Tonträger geriet die Schweizer Spieldosenindustrie in eine schwere Krise. Nach anfänglichem Zögern begann man schließlich auch in der Schweiz, billige Spieldosen mit Lochplatten zu produzieren. Ein Beispiel findet sich bei den plattengesteuerten Automaten.

Neben den extrem hochwertigen und teuren Spielwerken der Spieldosen wurden auch millionenfach billige Werke produziert, die u.a. in alle möglichen Alltagsgegenstände eingebaut wurden, vom Bierseidel bis zum Fotoalbum. In dieser Form haben die Spieluhren bis heute überlebt.

Music Boxes and Musical Movements

Besides pipes, metal combs belonged to the first sound sources used in automatic musical instruments. The metal comb as sound source was invented circa 1796 by the Swiss clock maker Antoine Favre in Geneva. The first movements with metal comb were used in clocks. In the following years a successful music box industry developed in Switzerland mainly located in the Geneva area. Simple musical movements with combs containing less than ten teeth developed into masterpieces of precision mechanics which were built into valuable cases and called music boxes. Besides the metal comb many music boxes contain further sound sources like little drums, bells or reeds. The most important manufacturer was Nicole Frères of Geneva. In Germany the Lamy Company in Furtwangen produced high-quality music boxes.

All classical music boxes use pinned barrels as sound carriers. The largest of them carry tens of thousands of pins which nonetheless is sufficient for a few pieces of music only. Therefore, with the invention of the cheap perforated metal disc as sound carrier the Swiss music box industry got into deep trouble. After brief hesitation Switzerland also started to produce cheap music boxes with perforated metal discs. One example is shown in the disc music box chapter of this book.

Besides the elaborated and very expensive movements of the music boxes, millions of cheap movements were produced that were built into objects of everyday use ranging from beer glasses to photo albums. In this style musical movements survived until today.

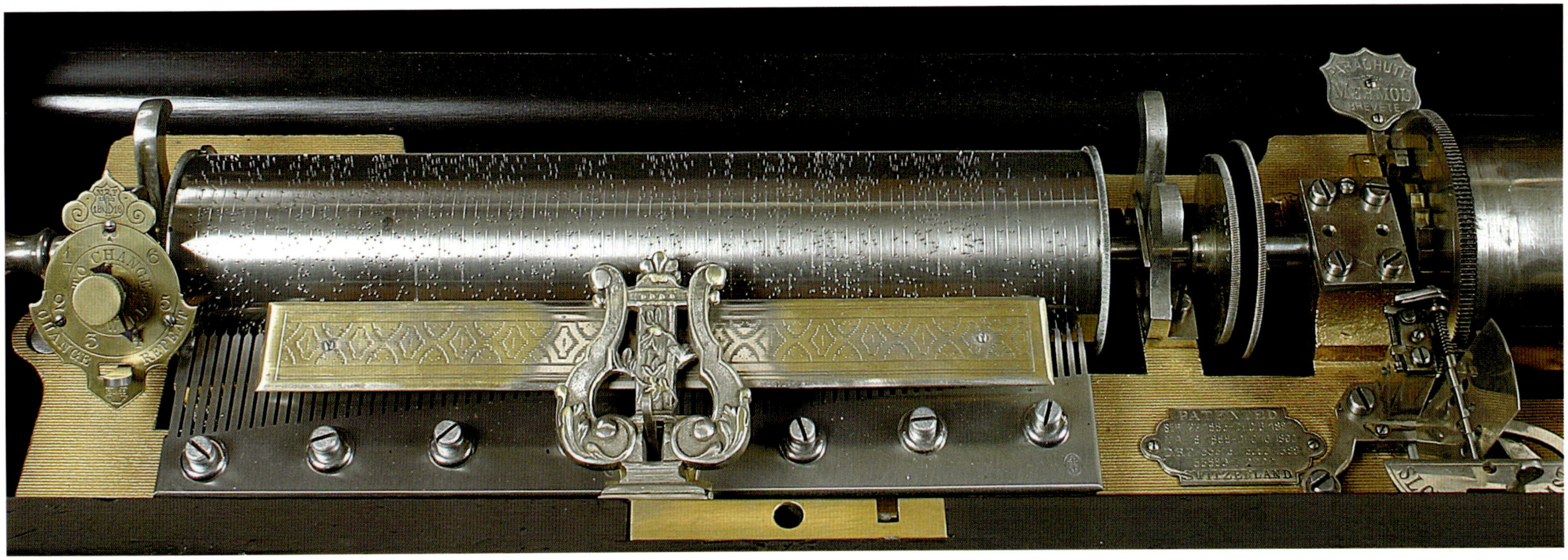

Schweizer Spieldose - Swiss Music Box

- Ungewöhnlich große, kunstvoll gefertigte Schweizer Spieldose mit Metallkamm, Glockenspiel und kleiner Trommel
- Gebaut um 1875 von B. A. Bremond / Genf
- In der Ablage unterhalb des Geräts befinden sich zwei Ersatzwalzen, Walzenbreite 65,5 cm
- Aufgrund der Größe erfolgt der Antrieb durch zwei Federmotoren
- Maße (HxBxT): 45 cm (mit Tisch 114 cm) x 140 cm x 60 cm

- Unusually large and elaborate Swiss music box with metal comb, chimes and a small drum
- Built around 1875 by B. A. Bremond / Geneva
- In the depository beneath the box there are two spare cylinders, barrel width 65.5 cm
- Due to its size the music box is driven by two spring-wound clockworks
- Dimensions (HxWxD): 45 cm (with table 114 cm) x 140 cm x 60 cm

Schweizer Spieldosen - Swiss Music Boxes

- Oben: Schweizer Spieldose mit Metallkamm, gebaut 1893 von J.H. Heller in Bern. Maße (HxBxT): 13 cm x 41 cm x 19 cm. Die Walze hat eine Breite von 21 cm und spielt sechs Musikstücke.
- Rechts: Sehr große, seltene Schweizer Spieldose mit zwei Metallkämmen, neun Glocken und Engelszungen (Harmonika-Zungen), gebaut vermutlich um 1880 von Baker & Co. in Genf. Im unteren Teil des Gehäuses befindet sich ein Gebläse, das die Luft für die Zungen liefert. Maße (HxBxT): 36,5 cm x 101 cm x 47,5 cm. Walzenbreite 54,5 cm.

- Top: Swiss music box with metal comb, built around 1893 by J.H. Heller in Bern. Dimensions (HxWxD): 13 cm x 41 cm x 19 cm. The barrel has a width of 21 cm and plays six pieces of music.
- Right: Very large and rare Swiss music box with two metal combs, nine bells, and "angels' tongues" (accordion reeds), probably built around 1880 by Baker & Co. in Geneva. In the lower part of the case there is a pump that produces the compressed air for the accordion reeds. Size (HxWxD): 36.5 cm x 101 cm x 47.5 cm. Barrel width 54.5 cm.

Schweizer Spieldose
Swiss Music Box

- Spieldose mit Metallkamm, gebaut um 1890 von Mermod in Sainte-Croix im Schweizer Jura
- Maße der Spieldose (HxBxT): 20 cm (mit Tisch ohne Aufsatz 99 cm) x 74 cm x 31 cm. Walzenbreite 28,5 cm.
- Fünf Ersatzwalzen vorhanden

- Music box with metal comb, built around 1890 by Mermod in Sainte-Croix in the Swiss Jura mountains
- Dimensions of the music box (HxWxD): 20 cm (with table excluding the top 99 cm) x 74 cm x 31 cm. Barrel width 28.5 cm.
- Five spare barrels available

Schweizer Spieldose - Swiss Music Box

- Spieldose aus Genf gebaut um 1880 mit Metallkamm und Glockenspiel
- Maße der Spieldose (HxBxT): 25 cm (mit Tisch 98,5 cm) x 77 cm x 36 cm. Walzenbreite 33 cm.

- Music box built around 1880 in Geneva with metal comb and chimes
- Size of the music box (HxWxD): 25 cm (with table 98.5 cm) x 77 cm x 36 cm. Barrel width 33 cm.

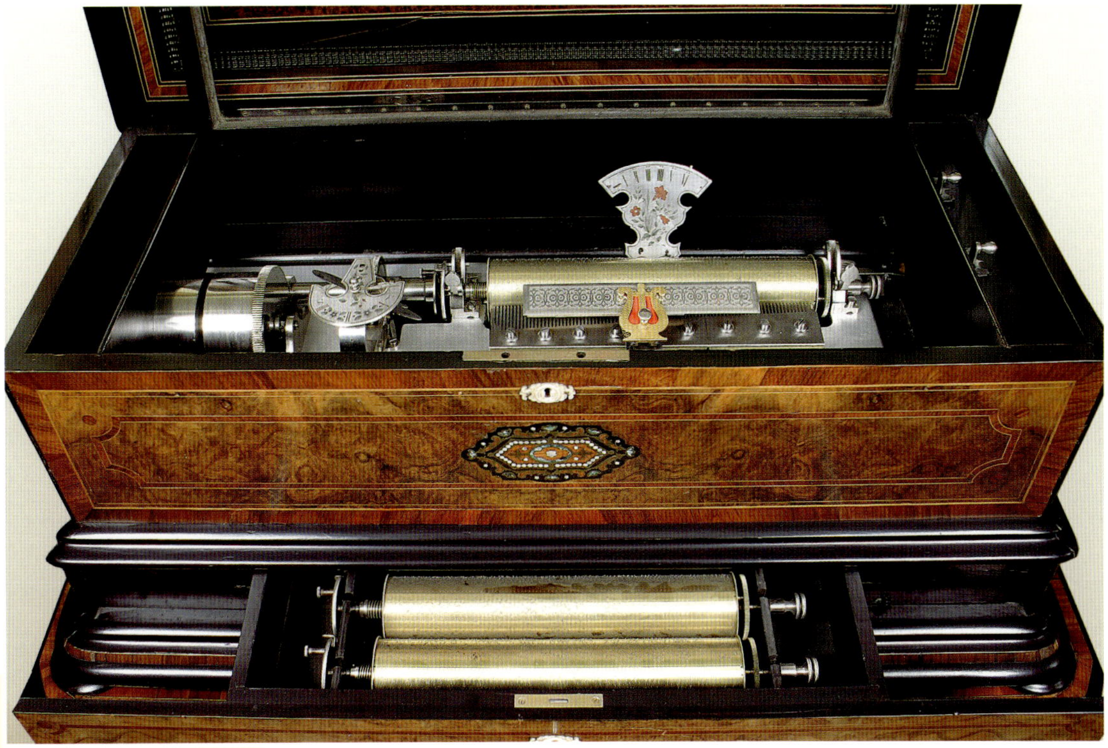

Schweizer Spieldose - Swiss Music Box

- Spieldose mit Metallkamm. Hersteller unbekannt, vermutlich gebaut in Sainte-Croix im Schweizer Jura.
- Maße der Spieldose (HxBxT): 43 cm (mit Tisch 117 cm) x 105 cm x 49 cm. Walzenbreite 33 cm.
- Zwei Ersatzwalzen vorhanden

- Music box with metal comb. Manufacturer unknown, probably built in Sainte-Croix in the Swiss Jura mountains
- Dimensions of the music box (HxWxD): 43 cm (with table 117 cm) x 105 cm x 49 cm. Barrel width 33 cm.
- Two spare barrels available

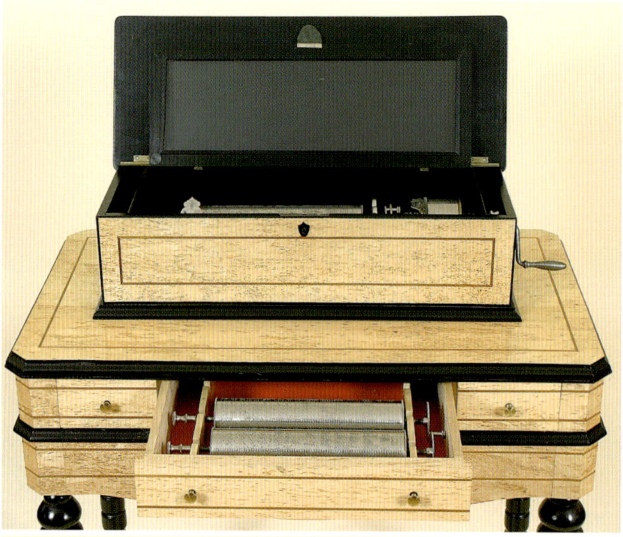

Schweizer Spieldose
Swiss Music Box

- Spieldose mit Metallkamm, gebaut um 1890 von Mermod in Sainte-Croix im Schweizer Jura
- Maße der Spieldose (HxBxT): 19,5 cm (mit Tisch 115 cm) x 71 cm x 28,5 cm. Walzenbreite 28,5 cm.
- Zwei Ersatzwalzen vorhanden

- Music box with metal comb, built around 1890 by Mermod in Sainte-Croix in the Swiss Jura mountains
- Dimensions of the music box (HxWxD): 19.5 cm (with table 115 cm) x 71 cm x 28.5 cm. Barrel width 28.5 cm.
- Two spare barrels available

Schweizer Spieldose - Swiss Music Box

- Spieldose mit Metallkamm, gebaut zwischen 1880 und 1890 von Karrer & Cie. in Teufenthal im Kanton Aargau (Deutschschweiz)
- Maße der Spieldose (HxBxT): 19 cm (mit Tisch 95 cm) x 77 cm x 30 cm. Walzenbreite 28 cm. Fünf Ersatzwalzen vorhanden

- Music box with metal comb built between 1880 and 1890 by Karrer & Cie. in Teufenthal in the canton Aargau (German-speaking part of Switzerland)
- Dimensions of the music box (HxWxD): 19 cm (with table 95 cm) x 77 cm x 30 cm. Barrel width 28 cm. Five spare barrels available.

Pfeifer-Figuren - Whistler Figures

- Oben: „Pfeifer" von Griesbaum in Triberg. Die Figur bewegt den Kopf und pfeift die Melodie „Muß i denn zum Städtele hinaus". Maße (HxBxT): 37 cm x 11 cm x 9 cm. Rechts: Uhrenträger um 1850. Die Figur bewegt Kopf, Augen und Mund und pfeift die Melodie „Oh du lieber Augustin". Wertvolles, funktionsfähiges Jockele Uhrwerk. Maße (HxBxT): 54 cm x 14 cm x 14 cm. Beide Figuren sind aus Holz geschnitzt und enthalten ein kleines, durch ein Federwerk angetriebenes Schöpfwerk, das die Luft für den Pfeifton liefert. Für einen Größenvergleich siehe das Vitrinenfoto im vorhergehenden Kapitel.

- Top: "Whistler" made by Griesbaum in Triberg. The figure moves its head while its playing. Dimensions (HxWxD): 37 cm x 11 cm x 9 cm. Right: Figure built around 1850 carrying a valuable clockwork made by Jockele. It moves its head, eyes, and mouth while its playing. Dimensions (HxWxD): 54 cm x 14 cm x 14 cm. Both figures are carved from wood and whistle melodies famous in Germany. They contain a small pump driven by a spring-wound clockwork that produces the air for the whistle tone. For size comparison see the picture of the showcase in the previous chapter.

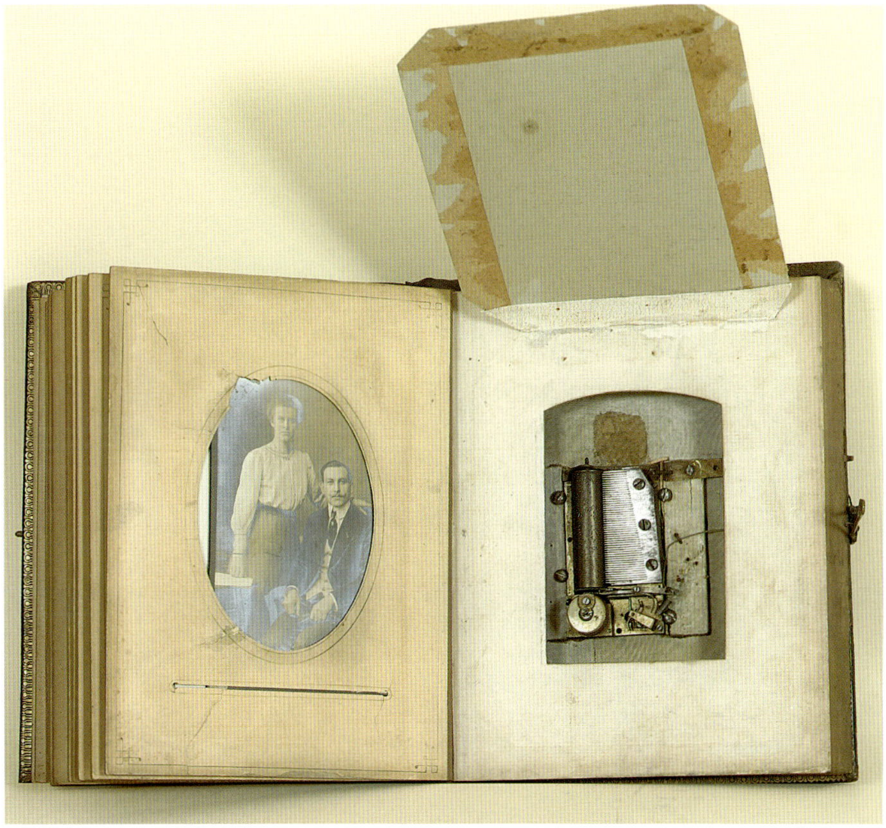

Fotoalbum mit Spieluhr
Photo Album with Musical Movement

- Fotoalbum mit Spieluhr um 1900, die beim Öffnen eine Melodie spielt. Größe (HxBxT): 27 cm x 22 cm x 7 cm.
- Photo album with musical movement from around 1900 which plays a melody when opened. Size (HxWxD): 27 cm x 22 cm x 7 cm.

Christbaumständer mit Spieluhr
Christmas Tree Stand with Musical Movement

- Christbaumständer mit Spieluhr und Glocken. Das Werk spielt zwei Melodien, gleichzeitig dreht sich der Baum.
- Christmas tree stand with musical movement and bells. The movement plays two melodies and turns the tree simultaneously.

EDISON
STANDARD
PHONOGRAPH

Grammophone und Jukeboxen

Ende der 1920er Jahren verschwanden die mechanischen Musikinstrumente mit Ausnahme der Kirmes- und Tanzorgeln fast völlig. Die Hauptgründe waren zum einen die sich anbahnende Weltwirtschaftskrise, zum anderen die zunehmende Verbreitung von Radioempfängern und elektrischen Plattenspielern. 1877 war es Edison erstmals gelungen, akustische Signale auf einer mit Zinnfolie überzogenen Walze aufzuzeichnen. 1887 erfand Emil Berliner die Schallplatte, die mittels Trichtergrammophon abgespielt werden konnte. 1907 erschien die Triodenröhre, mit der elektrische Verstärker möglich wurden. Um 1925 wurden die Trichtergrammophone von Plattenspielern mit elektrischen Verstärkern abgelöst. Musik war endgültig zu einem preiswerten Konsumartikel geworden, den sich jeder leisten konnte und der in nahezu beliebiger Vielfalt jederzeit verfügbar war.

Selbst die renommiertesten Hersteller von Musikautomaten konnten sich diesen Umwälzungen nicht entziehen. 1903 gab es allein in Deutschland 104 Produzenten. 1925 war kaum mehr als eine Handvoll davon übrig geblieben. Eine der ganz wenigen Firmen, die den Übergang zur neuen Technik bewältigte, war Wurlitzer in den USA. Hier baute man nach dem Zweiten Weltkrieg farbenprächtige Jukeboxen mit außergewöhnlichem Design, die perfekt dem Lebensgefühl der fünfziger Jahre entsprachen. Heute sind auch diese Geräte bereits wieder Klassiker, die bei Sammlern heiß begehrt sind.

Gramophones and Jukeboxes

At the end of the 1920s, with the exception of dance and fairground organs, automatic musical instruments almost completely disappeared. The main reasons were the beginning of the Great Depression and the wide-spread use of radio sets and electric record players. In 1877, Edison first succeeded in recording audio signals on a barrel covered with tin foil. In 1887, Emil Berliner invented the record which could be played with gramophones. In 1907, the triode tube appeared providing the basis for electric amplifiers. Around 1925, the gramophones were replaced by record players with electric amplifiers. By this time music eventually had become a cheap consumer good that was affordable for everyone and available in almost every style and variety.

Even the most renowned manufacturers of automatic musical instruments could not escape from this technical revolution. In 1903 there were 104 manufacturers in Germany alone. Only few survived until 1925. One of the very few companies which managed the transition to the new technology was Wurlitzer in the USA. After WW II Wurlitzer built colorful jukeboxes with an extravagant design which perfectly matched the lifestyle of the 1950s. Today these gadgets have already become classics which are highly desired by collectors.

Gegenüberliegende Seite: Edison Standard Phonograph von 1898. Oben links: Emil Berliner, der Erfinder der Schallplatte. Oben rechts: T.A. Edison im Alter von 31 Jahren mit seinem Phonographen.

Opposite page: Edison Standard Phonograph from 1898. Top left: Emil Berliner, inventor of the record. Top right: T.A Edison at the age of 31 with his phonograph.

Grammophone

- Oben: Edison Grammophon mit Tulpentrichter von 1905. Maße (HxBxT): 37 cm x 38 cm x 38 cm.
- Rechts: Koffergrammophon um 1940. Maße (HxBxT): 13,5 cm x 22 cm x 27,5 cm.

- Top: Edison gramophone with tulip like funnel from 1905. Dimensions (HxWxD): 37 cm x 38 cm x 38 cm.
- Right: Box gramophone built around 1940. Dimensions (HxWxD): 13,5 cm x 22 cm x 27,5 cm.

Grammophone

- Oben: Kinder-Phonograph der Fa. Rex um 1930. Maße (HxBxT): 34 cm x 36 cm x 23 cm.
- Links: Großgrammophon um 1930 mit riesigem Trichter aus Pappe, Durchmesser 73 cm, speziell für Gaststätten und Ballsäle. Doppelfederaufzug für besonders lange Spieldauer. Maße (HxBxT): 150 cm x 50 cm x 45 cm.

- Top: Phonograph for children built by the Rex company around 1930. Dimensions (HxWxD): 34 cm x 36 cm x 23 cm.
- Left: Large gramophone built around 1930 with a huge cardboard funnel, diameter 73 cm, specially for ball rooms and restaurants. The spring-wound clockwork with two springs allows for an extended playing time. Dimensions (HxWxD): 150 cm x 50 cm x 23 cm.

Oben und rechts: Das „50er-Jahre-Zimmer" im Wilhelmsbau. Vom Petticoat bis zur Wurlitzer-Jukebox findet sich hier alles, was den Flair der Zeit ausmachte, als der Rock'n Roll die Welt eroberte und Elvis die Hüften schwang.

Top and right: The "Fifties' Room" in the Wilhelmsbau. From petticoats to Wurlitzer jukeboxes everything can be found here that was typical for the time when the Rock'n'Roll conquered the world and Elvis gyrated his hips.

Wurlitzer Jukebox 1015

- Bauzeit 1946 - 1947
- Maße (HxBxT): 152 cm x 85 cm x 63,5 cm
- Mit über 56 000 verkauften Exemplaren war das Modell 1015 die erfolgreichste Wurlitzer Jukebox. Ein wichtiges Designelement waren acht beleuchtete Röhren, in denen Luftblasen aufstiegen. Daher wurde diese Art von Jukebox auch als „Bubbler" bezeichnet. Erfunden wurden sie vom emigrierten Schweizer Paul Fuller, der von 1934 bis 1948 Chefdesigner bei Wurlitzer war.

- Production period 1946 - 1947
- Dimensions (HxWxD): 152 cm x 85 cm x 63,5 cm
- With more than 56,000 units sold, the model 1015 was the most successful Wurlitzer jukebox. An important design element were eight "bubble tubes", illuminated glass tubes with a never-ending supply of rising bubbles. Therefore, this type of jukebox was also called "bubbler". It was invented by the emigrated Swiss Paul Fuller who was Wurlitzer's chief designer between 1934 - 1948.

Wurlitzer Jukebox 850 (links / left)

- Baujahr: 1941; Maße (HxBxT): 166 cm x 99 cm x 67 cm.
- Eine sehr teure Jukebox, damals wie heute. Sie trägt auch den Spitznamen „Peacock" („Pfau"). Eine Polaroid-Beleuchtung des Fensters mit den Pfauen sorgt für ständig wechselnde Farben.

- Built: 1941; Dimensions (HxWxD): 166 cm x 99 cm x 67 cm.
- A very expensive jukebox, then and today. It is nicknamed „Peacock". Polaroid illumination provides ever-changing colors on the peacock glass.

Wurlitzer Jukebox 1100 (rechts / right)

- Bauzeit: 1947 - 1949; Maße (HxBxT): 147 cm x 77,5 cm x 69 cm.
- Das erste Modell, bei dem durchsichtige Kunststoffteile verwendet wurden und das letzte, das von Paul Fuller gestaltet wurde.

- Production period: 1947 - 1949; Dimensions (HxWxD): 147 cm x 77,5 cm x 69 cm.
- The first model introducing clear plastics and the last one designed by Paul Fuller.

Musikautomaten im
Auto & Technik Museum Sinsheim

~•~

Automatic Musical Instruments in the
Auto & Technik Museum Sinsheim

Inhalt - Contents

 Fotografie, Texte, Layout und CD-Mastering: Dr. Hans-Jürgen Schlicht Multimedia & Print, Sinsheim. Übersetzung: Dr. Hans-Jürgen Schlicht und Bianca Sowders. Unser herzlicher Dank gilt Gotthard Arnold, Helga Erbacher und Sina Hildebrand von der Fachstätte historischer Musikautomaten in Bad Schönborn für die tatkräftige Unterstützung.

ISBN 3-9809437-4-7

Die mechanischen Musikinstrumente sind ein interessanter Teil der Technikgeschichte. Die umfangreiche Sammlung, die heute in den Museen Sinsheim und Speyer besichtigt werden kann, hat ihren Ursprung im Auto & Technik Museum Sinsheim. Schon bei der Eröffnung im Jahr 1981 wurden in den Museumshallen in Sinsheim zahlreiche Musikautomaten gezeigt. Heute sind die meisten Instrumente im Wilhelmsbau im Technik Museum Speyer ausgestellt. Neben einer Reihe kleinerer Musikautomaten befinden sich aber zahlreiche bedeutende Großorgeln, wie z.B. die von der Firma Mortier gebaute größte Tanzorgel der Welt und eine Aeolian Grand Konzertorgel mit mehreren tausend Pfeifen, nach wie vor in Sinsheim.

Die meisten Instrumente sind voll spielfähig und können von den Besuchern in Gang gesetzt oder im Rahmen von Führungen in Aktion erlebt werden. Einen Eindruck von der Klangqualität gibt die diesem Buch beiliegende Audio-CD. Bei den Instrumenten ist vermerkt, wenn sich ein Musikbeispiel auf der CD befindet.

Automatic musical instruments represent an interesting part of technical history. The extensive collection which can be visited in the museums in Sinsheim and Speyer originated in the Auto & Technik Museum Sinsheim. At its opening in 1981, many automatic instruments were on display in the museum halls in Sinsheim already. Today most instruments are shown in the Wilhelmsbau in the Technik Museum Speyer. However, besides numerous smaller instruments many important large organs like the biggest dance organ in the world, built by the Mortier company, and an Aeolian Grand concert organ with several thousand pipes are still in Sinsheim.

Most of the instruments are fully playable and can either be activated by the visitors or can be experienced during a guided tour. The audio CD in this book gives an impression of their quality of sound. You will find a note at the instrument description if a music example is contained on the CD.

GEBROEDERS DECAP ANTWERPEN
GEBR. DECAP
GEBR. DECAP
ORGAN JAZZ

Das Auto & Technik Museum Sinsheim

- Eröffnet am 6. Mai 1981
- Überdachte Ausstellungsfläche mit über 30 000 m² in zwei Hallen
- Großes Freigelände
- Tausende Exponate aus allen Bereichen der Technikgeschichte
- Große Sammlung mechanischer Musikinstrumente, darunter die größte Tanzorgel der Welt
- „Concorde" der Air France und „Russische Concorde" Tupolev 144
- Weltweit größte Privatsammlung von Mercedes-Kompressor-Automobilen
- Größte Sammlung von Maybach- und Bugatti-Automobilen in Deutschland
- Über 300 weitere Oldtimer, 200 Motorräder, 100 Renn- und Sportwagen, 22 Lokomotiven, 60 Flugzeuge, 150 Traktoren, Dampfmaschinen und LKWs, historische Moden u.v.m.
- Militärhistorische Abteilung
- Größte permanente Formel-1-Ausstellung in Europa
- Über 300 Dragster, Straßenkreuzer, Rallye- und Spezialfahrzeuge

The Auto & Technik Museum Sinsheim

- Opened on 6. May 1981
- Covered exhibition area with more than 30,000 sqm in two halls
- Large open air ground
- Thousands of exhibits from all fields of technical history
- Large collection of automatic musical instruments, including the largest dance organ in the world
- "Concorde" of Air France and "Russian Concorde" Tupolev 144
- Largest private collection of Mercedes cars with supercharged engine worldwide
- Largest collection of Maybach- and Bugatti cars in Germany
- More than 300 additional vintage cars, 200 motorcycles, 100 racing- and sports cars, 22 locomotives, 60 airplanes, 150 tractors, steam engines and trucks, historical fashion and much more
- Department of military history
- Largest permanent Formula 1 exhibition in Europe
- More than 300 dragsters, American dream cars, rallye- and special utility vehicles

Audio CD Titel 32 + 33

Die größte Tanzorgel der Welt
The Largest Dance Organ in the World

- Gebaut von Mortier in Antwerpen / Belgien, einem der bedeutendsten Hersteller von mechanischen Großorgeln
- Pneumatiksteuerung mit elektrischem Antrieb, spielt Kartonnoten mit einer Breite von 435 mm
- Maße (HxBxT): 7 Meter x 8 Meter x 4,4 Meter
- Die Tanzorgel belegt eine Grundfläche von über 30 m^2
- Das Grundinstrument entstand um 1912
- In der jetzigen Ausführung verfügt die Tanzorgel über 900 Pfeifen, ein Saxophon, zwei Akkordeons mit 41 Pianotasten und 120 Bässen (chromatisch) sowie eine vollständige Schlagzeugausrüstung. Sie spielt die gesamte Palette der Tanzmusik sowie klassische Werke.
- Über 1000 Kartonbänder, die in der ganzen Welt gesammelt wurden, vorhanden

- Built by Mortier in Antwerp / Belgium, one of the most important manufacturers of self playing large organs
- Pneumatic control with electro motor, plays music books with a width of 435 mm
- Dimensions (HxWxD): 7 Meter x 8 Meter x 4,4 Meter
- The organ covers a surface area of 30 sqm
- The basic instrument was built around 1912
- In its current form the dance organ contains 900 pipes, one saxophone, two accordions with 41 piano keys and 120 bass keys (chromatic) and a complete percussion part. It plays the whole range of dance music as well as classical music.
- More than 1,000 music books, which were collected in the whole world, available

MORTIER
ROCK'N ROLL JAZZ
DE NACHTEGAAL

Gegenüberliegende Seite: Die Tanzorgel verfügt über eine herrliche Beleuchtung. Bilder oben und rechts: Die Steuerung der Orgel erfolgt pneumatisch, als Tonträger dient ein perforiertes Kartonband. Solche Kartonbänder werden auch als Notenbücher bezeichnet, da sie in Form eines Buches gefaltet werden. Für die Mortier Tanzorgel sind über 1000 Notenbücher vorhanden.

Opposite page: The organ has a beautiful illumination. Pictures top and right: The organ is equipped with pneumatic control. It uses a perforated cardboard band as sound carrier. Such cardboard bands are also referred to as "music books" since they are folded zig-zag fashion into a compact pile that has the form of a book. More than 1,000 music books are available for the Mortier dance organ.

Audio CD Titel 34

Hooghuys Orgel - Hooghuys Organ

- Karussellorgel mit 84 Tonstufen und Schwellwerk mit weltweit einzigartiger Fassade
- Gebaut um 1910 von Hooghuys in Willebroek / Belgien
- Die Firma Hooghuys wurde 1880 von Louis und Edmond Hooghuys gegründet und war neben Mortier und Decap eine der wichtigsten Firmen des klassischen Großorgelbaus
- Pneumatiksteuerung mit elektrischem Antrieb, spielt Kartonnoten mit einer Breite von 315 mm
- Maße (HxBxT): 4,30 Meter x 7 Meter x 1,70 Meter
- Instrumente: Orgelpfeifen, Xylophon, Schlagzeug, Piccolo, vox humana

- Carousel organ with 84 keys and swell shutters with a worldwide unique facade
- Built around 1910 by Hooghuys in Willebroek / Belgium
- The Hooghuys company was founded in 1880 by Louis and Edmond Hooghuys and was, besides Mortier and Decap, one of the most important manufacturer of classic large organs
- Pneumatic control with electro motor, plays music books with a width of 315 mm
- Dimensions (HxWxD): 4.30 Meter x 7 Meter x 1.70 Meter
- Instruments: Organ pipes, xylophone, percussion, piccolo, vox humana

ARNOLD

Aeolian Grand Konzertorgel
Aeolian Grand Concert Organ

- Größte selbstspielende Orgel der Welt. Papierrolle mit 260 mm Breite als Tonträger. Die Orgel kann auch manuell gespielt werden.
- Das als Konzert- und Kathedralorgel gedachte Instrument wurde von der Schweizer Orgelbauanstalt Gatringer in den Jahren zwischen 1928 und 1930 gebaut. Der mechanische Steuerteil stammt von der amerikanischen Firma Aeolian.
- Die Orgel wurde mit grober Gewalt abgebaut und von Herrn Dr. Weiss-Winkler vor der totalen Zerstörung bewahrt. Nach einer Zwischenlagerung in Solothurn gelangte sie Mitte der 1980er Jahre nach Sinsheim, wo sie von der Firma Arnold umfassend restauriert wurde.
- Maße (HxBxT): 7,20 Meter x 8,60 Meter x 5,60 Meter
- Die größte Pfeife ist 8,5 Meter lang (gekröpft eingebaut), wiegt ca. 435 Kg und erzeugt einen akustisch nicht mehr wahrnehmbaren Ton mit einer Frequenz von nur 16 Hz. Die kleinste Pfeife ist ein Winzling von 8 mm Länge und 10 Gramm Gewicht. Sie erzeugt einen gerade noch hörbaren Ton von 18 000 Hz.
- Die Orgel verfügt insgesamt über mehr als 4000 Pfeifen, von denen bis jetzt ca. 2500 eingebaut werden konnten
- Durch die vollautomatische Steuerung kann die Orgel so gespielt werden, als hätte der Organist 60 Finger
- Die mehr als 1000 Musikrollen, die für die Aeolian Grand vorhanden sind, umfassen die gesamte Palette der modernen und klassischen Musik

- Largest self-playing organ in the world. Paper roll with a width of 260 mm as sound carrier. The organ can also be played manually.
- The instrument that was intended as concert- or cathedral organ was built between 1928 and 1930 by the Swiss organ builder Gatringer. The mechanic control was built by the American company Aeolian.
- The organ was disassembled with brute force and saved from total destruction by Mr. Dr. Weiss-Winkler. After temporary storage in Solothurn it was transported to Sinsheim where it was completely restored by Messrs. Arnold in the mid 1980s.
- Dimensions (HxWxD): 7,20 Meter x 8,60 Meter x 5,60 Meter
- The largest pipe is 8.5 meters long (built in bended), weighs about 435 kg and produces an acoustical not perceivable sound with a frequency of 16 Hz. The smallest pipe is a midget with a length of 8 mm and weighs a mere 10 grams. It generates a just perceivable sound of 18,000 Hz.
- The organ has more than 4,000 pipes of which 2,500 could be installed so far
- Due to the fully automatic control the organ can play like an organist with 60 fingers
- The more than 1000 music rolls that are available for the aeolian grand comprise the whole range of modern and classical music

Oben: Für die Orgel sind über 1000 originale Musikrollen vorhanden. Rechts: Eine der riesigen Orgelpfeifen bei der Montage.

Top: There are more than 1000 music rolls available for the organ. Right: One of the huge organ pipes during assembly.

Heinkel HE 111 H
AEOLIEN

Gegenüberliegende Seite: Die Aeolian Grand wurde auf einem Podest in der Halle 1 des Museums aufgestellt, das von unten einen Blick auf die Orgelmechanik ermöglicht. Links: Die längste Pfeife hat eine Länge von 8,5 Metern und wurde gekröpft eingebaut. Oben: Der riesige Balg des Schöpfwerks umgeben von einem Teil der 2500 Pfeifen.

Opposite page: The Aeolian Grand was placed on a platform in hall 1 of the Museum Sinsheim that allows to watch the mechanics of the organ from below. Left: The largest pipe has a length of 8,5 meters and was built in bended. Top: The huge bellow of the air pump surrounded by some of the more than 2,500 pipes.

Decap Tanzorgel mit Drehbühne - Decap Dance Organ with Revolving Platform

- Gebaut um 1930 von Decap in Antwerpen / Belgien
- Pneumatiksteuerung mit elektrischem Antrieb, spielt Kartonnoten mit einer Breite von 325 mm
- Maße (HxBxT): 2,60 Meter x 3,60 Meter x 1,65 Meter
- Instrumente: Zwei Hohner-Akkordeons mit 41 Pianotasten und 120 Bässen (chromatisch), ein Saxophon, Schlagzeug

- Built around 1930 by Decap in Antwerp / Belgium
- Pneumatic control with electro motor, plays music books with a width of 325 mm
- Dimensions (HxWxD): 2.60 Meter x 3.60 Meter x 1.65 Meter
- Instruments: Two Hohner accordions with 41 piano keys and 120 bass keys (chromatic), one saxophone, percussion

Decap Tanzorgel - Decap Dance Organ

- Gebaut 1953 von Decap in Antwerpen / Belgien
- Pneumatiksteuerung mit elektrischem Antrieb, spielt Kartonnoten Breite 325 mm, Maße (HxBxT): 2,70 m x 4,20 m x 1,50 m
- Instrumente: Zwei Akkordeons mit 41 Pianotasten und 120 Bässen (chromatisch), zwei Saxophone, Schlagzeug, Holzpfeifen und elektronische Klangerzeugung mit Radioröhren

- Built around 1953 by Decap in Antwerp / Belgium
- Pneumatic control with electro motor, plays music books width 325 mm, Dimensions (HxWxD): 2.70 m x 4.20 m x 1.50 m
- Instruments: Two accordions with 41 piano keys and 120 bass keys (chromatic), two saxophones, percussion, wooden pipes, and electronic sound generation with radio tubes

DECAP

Decap Tanzorgel
Decap Dance Organ

- Tanzorgel mit Drehbühne gebaut um 1934 von Decap in Antwerpen / Belgien
- Pneumatiksteuerung mit elektrischem Antrieb, spielt Kartonnoten mit einer Breite von 330 mm
- Maße (HxBxT): 2,80 m x 4,10 m x 1,60 m
- Instrumente: Ein Hohner-Akkordeon mit 41 Pianotasten und 120 Bässen (chromatisch), ein Saxophon, vier Reihen Holzpfeifen, Schlagzeug

- Dance organ with revolving platform built around 1934 by Decap in Antwerp / Belgium
- Pneumatic control with electro motor, plays music books with a width of 330 mm
- Dimensions (HxWxD): 2.80 m x 4.10 m x 1.60 m
- Instruments: Two Hohner accordions with 41 piano keys and 120 bass keys (chromatic), one saxophone, four rows of wooden pipes, percussion

Beim Spielen dreht sich das Podest mit der Tanzorgel. Dadurch sehen die Besucher nicht nur die wunderschöne Fassade, sondern auch die beeindruckende Mechanik des Instruments.

The platform revolves while the dance organ is playing. That way the visitors not only see the beautiful facade but also the impressive mechanic control of the instrument.

PHILIPPS - VIOLINE
1919

PHILIPPS - VIOLINE
1919

Philipps Violine

- Gebaut 1919 von Philipps in Frankfurt a. M., ergänzt 2005 von der Firma Arnold in Bad Schönborn
- Höchstes Orchestrion der Welt
- Pneumatiksteuerung mit elektrischem Antrieb, perforierte Papierrolle mit 226 mm Breite als Tonträger
- Maße (HxBxT): 4,50 m x 2,80 m x 1,10 m
- Gehäuse aus massivem Eichenholz mit reichhaltigen Schnitzereien, Facettenspiegel mit aufgesetzten Emblemen, gläserne Schwellerjalousien
- Instrumente: Klavier, Mandoline, mit Betonung spielendes Xylophon, 4' und 8' Violinenpfeifenreihen, Akkordeon mit 120 Bässen, Piccolo, Vox humana, Kastagnetten, Triangel, Trommel und Pauke
- Aufwändige Beleuchtung mit wechselnden Lichteffekten

Philipps Violin

- Built 1919 by Philipps in Frankfurt a. M., supplemented 2005 by Messrs. Arnold in Bad Schönborn
- Highest Orchestrion in the world
- Pneumatic control with electro motor, perforated paper roll with a width of 226 mm as sound carrier
- Dimensions (HxWxD): 4.50 m x 2.80 m x 1.10 m
- Cabinet made of massive oak wood with extensive carvings, facet mirrors with emblems, swell shutters made of glass
- Instruments: Piano, mandoline, xylophone with expression, 4' and 8' violin pipe rows, accordion with 120 bass keys, piccolo, vox humana, castanets, triangle, drum and bass drum
- Elaborate illumination with changing light effects

Oben: Der mit vielen Schnitzereien verzierte Instrumentenaufsatz mit den 8' Violinenpfeifen und dem Akkordeon. Rechts: Im Mittelteil befinden sich neben der Steuerung u.a. das Xylophon, die kleine Trommel und die 4' Violinenpfeifen.

Top: The instrument top, decorated with many carvings, with the 8' violin pipes and the accordion. Right: Besides the control the middle part contains the xylophone, the small drum, and the 4' violin pipes.

Popper „Bianca" Orchestrion

- Gebaut um 1915 von Popper in Leipzig
- Pneumatiksteuerung mit elektrischem Antrieb, perforierte Papierrolle mit 345 mm Breite als Tonträger
- Maße (HxBxT): 260 cm (mit Aufbau 400 cm) x 155 cm x 100 cm
- Instrumente: Klavier, Mandoline, Xylophon, Glockenspiel und Schlagzeug
- Vom „Bianca" gibt es viele unterschiedliche Gehäusevariante. Es wurde üblicherweise in Gaststätten und Tanzcafés aufgestellt.

- Built around 1915 by Popper in Leipzig
- Pneumatic control with electro motor, perforated paper roll with a width of 345 mm as sound carrier
- Dimensions (HxWxD): 260 cm (400 cm with top) x 155 cm x 100 cm
- Instruments: Piano, mandoline, xylophone, chimes, and percussion
- The "Bianca" was built in many different cabinet styles. It was usually placed in guesthouses, pubs and dance cafes.

Popper „Happy Jazzband“

- Gebaut um 1920 von Popper in Leipzig
- Pneumatiksteuerung mit elektrischem Antrieb, perforierte Papierrolle mit 345 mm Breite als Tonträger
- Maße (HxBxT): 230 cm x 155 cm x 75 cm
- Instrumente: Klavier, Mandoline und Schlagzeug

- Built around 1920 by Popper in Leipzig
- Pneumatic control with electro motor, perforated paper roll with a width of 345 mm as sound carrier
- Dimensions (HxWxD): 230 cm x 155 cm x 75 cm
- Instruments: Piano, mandoline, percussion

Audio CD Titel 35

Automatisches Klavier von Popper - Popper Reproducing Piano

- Gebaut um 1915 von der Firma Popper in Leipzig
- Pneumatiksteuerung mit elektrischem Antrieb, perforierte Papierrolle mit 345 mm Breite als Tonträger
- Maße (HxBxT): 150 cm x 160 cm x 75 cm

- Built around 1915 by Popper in Leipzig
- Pneumatic control with electro motor, perforated paper roll with a width of 345 mm as sound carrier
- Dimensions (HxWxD): 150 cm x 160 cm x 75 cm

Lochmann „Original“

- Plattengesteuerter Musikautomat mit Federantrieb und Handaufzug
- Gebaut um 1900 von Lochmann in Leipzig
- Maße (HxBxT): 139 cm x 78 cm x 41 cm (mit Unterschrank 232 cm x 90 cm x 45 cm); Plattendurchmesser 62 cm
- Instrumente: Zwei Doppelkämme aus Metall, 12 Röhrenglocken, Mandolineneffekt, Schlaginstrumente

- Disc music box with spring-wound clockwork
- Built around 1900 by Lochmann in Leipzig
- Dimensions (HxWxD): 139 cm x 78 cm x 41 cm (with lower cabinet 232 cm x 90 cm x 45 cm); disc diameter 62 cm
- Instruments: Two metal double combs, 12 tubular chimes, mandoline effect, percussion

Polyphon

- Gebaut 1895 von dem Polyphon Musikwerken in Leipzig
- Plattengesteuerter Musikautomat mit Federantrieb und Metallkamm
- Plattendurchmesser 62,5 cm
- Maße (HxBxT): 255 cm x 90 cm x 50 cm
- Das Instrument wurde anlässlich des 80. Geburtstags des Reichskanzlers Bismarck mit einer Plakette geschmückt mit der Inschrift: „Wir fürchten Gott, sonst nichts in der Welt. 1. April 1895".

- Built around 1895 by the Polyphon Musikwerke in Leipzig
- Disc music box with spring-wound clockwork and metal comb
- Disc diameter 62.5 cm
- Dimensions (HxWxD): 255 cm x 90 cm x 50 cm
- On occasion of the 80th birthday of then German chancellor Bismarck, the instrument was decorated with a badge with the inscription: "We fear God and nothing else in the world. 1. April 1895".

Kirmesorgel
Fairground Organ

- Kirmesorgel mit Originalfassade von Giovanni Bacigalupo / Berlin
- Maße (HxBxT): 330 cm x 280 cm x 125 cm
- Orgelteil mit drei Manualen mit 40 Registern plus Bassregister

- Fairground organ with original facade by Giovanni Bacigalupo / Berlin
- Dimensions (HxWxD): 330 cm x 280 cm x 125 cm
- Organ part with three manuals with 40 stops plus bass stop

Molzer Kirmesorgel - Molzer Fairground Organ

- Gebaut um 1926 von Molzer in Wien
- Pneumatiksteuerung mit elektrischem Antrieb, spielt Kartonnoten mit 58er Skala
- Maße (HxBxT): 1,8 m x 2,4 m x 0,85 m

- Built around 1926 by Molzer in Vienna
- Pneumatic control with electro motor, plays 88-key music books
- Dimensions (HxWxD): 1,8 m x 2,4 m x 0.85 m

Weiss Kirmesorgel - Weiss Fairground Organ

- Gebaut 1982 von Weiss in Freisen / Niederlande
- Pneumatiksteuerung mit elektrischem Antrieb, Lochkartonband als Tonträger
- Maße (HxBxT): 2 m x 3 m x 0,8 m

- Built in 1982 by Weiss in Freisen / The Netherlands
- Pneumatic control with electro motor, plays music books
- Dimensions (HxWxD): 2 m x 3 m x 0.8 m

Drehorgel von Giovanni Bacigalupo

- Eine der größten fahrbaren, handbetriebenen Drehorgeln aus der Produktion der Firma Bacigalupo, gebaut 1893 in Berlin
- Maße (HxBxT): 190 cm x 140 cm x 80 cm
- 167 Pfeifen, angeordnet in neun Registern, davon vier Zungenregister und ein Piccolo
- Gesteuert von einer hölzernen Stiftwalze mit einer Breite von 110 cm spielt das Instrument zehn Melodien

Hand-cranked Organ by Giovanni Bacigalupo

- One of the largest mobile hand-cranked organs made by Messrs. Bacigalupo, built 1893 in Berlin
- Dimensions (HxWxD): 190cm x 140cm x 80cm
- 167 pipes, nine stops including four reed stops and one piccolo
- The pinned barrel has a width of 110 cm and plays ten tunes

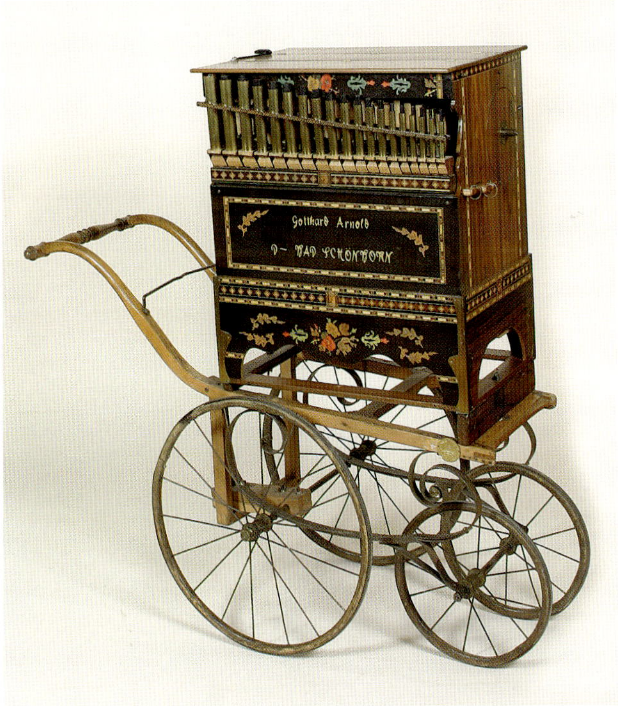

Drehorgeln Hand-cranked Organs

- Links: Drehorgel mit 25 Tonstufen mit einem Grundregister und einer Pfeifenreihe gebaut 1975 von der Fa. Arnold unter Verwendung alter Originalteile. Maße (HxBxT): 106 cm x 48 cm x 27 cm.
- Rechts: Walzendrehorgel mit drei Registern mit 33 Tonstufen gebaut 1980 von der Fa. Arnold. Maße (HxBxT): 123cm x 58cm x 40cm.

- Left: Hand-cranked organ with 25 keys with one basic stop and a row of pipes built in 1975 by Messrs. Arnold using old original parts. Dimensions (HxWxD): 106cm x 48cm x 27cm.
- Right: Barrel organ with three stops and 33 keys built in 1980 by Messrs. Arnold. Dimensions (HxWxD): 123cm x 58cm x 40cm.

Drehorgeln - Hand-cranked Organs

- Oben links: Violinopan Walzendrehorgel mit 26 Tonstufen mit Pfeifen und Violinpfeifen gebaut im letzten Viertel des 19. Jahrhunderts von G. Bacigalupo in Berlin. Renovation 1981 durch Arnold in Bad Schönborn. Maße (HxBxT): 123 cm x 50 cm x 32 cm.
- Oben rechts: Walzendrehorgel mit 25 Tonstufen und Originalwalze gebaut 1893 von F. Kolb & Söhne. Renovation 2004 durch Arnold in Bad Schönborn. Maße (HxBxT): 115 cm x 47 cm x 31 cm.
- Unten links: 38er Drehorgel gebaut 1900 von Cocchi, Bacigalupo & Graffigna in Berlin. Maße (HxBxT): 115 cm x 64 cm x 37 cm.

- Top left: Hand-cranked violinopan barrel organ with 26 keys with pipes and violin pipes built in the last quarter of the 19th century by G. Bacigalupo in Berlin. Renovated 1981 by Messrs. Arnold / Bad Schönborn. Dimensions (HxWxD): 123 cm x 50 cm x 32 cm.
- Top right: Hand-cranked barrel organ with 25 keys and original barrel built 1893 by F. Kolb & Söhne. Renovated 2004 by Messrs. Arnold / Bad Schönborn. Dimensions (HxWxD): 115 cm x 47 cm x 31 cm.
- Bottom left: Hand-cranked organ with 38 keys built in 1900 by Cocchi, Bacigalupo & Graffigna in Berlin. Dimensions (HxWxD): 115 cm x 64 cm x 37 cm.